Location, Localization, and Localizability

Yunhao Liu • Zheng Yang

Location, Localization, and Localizability

Location-awareness Technology for Wireless Networks

Yunhao Liu
Tsinghua University
Beijing, China
yunhao@greenorbs.com

Zheng Yang
Tsinghua University
Beijing, China
yang@greenorbs.com

ISBN 978-1-4899-8560-6 ISBN 978-1-4419-7371-9 (eBook)
DOI 10.1007/978-1-4419-7371-9
Springer New York Dordrecht Heidelberg London

Softcover re-print of the Hardcover 1st edition 2011

Printed on acid-free paper

Springer is part of Springer Science+Business Media (www.springer.com)

About the Authors

Yunhao Liu received his B.S. degree in Automation Department from Tsinghua University, China, in 1995, an M.A. degree in Beijing Foreign Studies University, China, in 1997, and an M.S. and a Ph.D. degree in Computer Science and Engineering at Michigan State University in 2003 and 2004, respectively. He is a professor at Tsinghua University, as well as a distinguished researcher at Tsinghua National Lab for Information Science and Technology. He is also a faculty member at the Hong Kong University of Science and Technology. He serves as Associate Editor for IEEE Transactions on Parallel and Distributed Systems and IEEE Transactions on Mobile Computing. He also served as the general vice-chair of WWW 2008, an area chair of ICDCS 2010, the PC Co-Chair of MASS 2011, ICPADS 2011, and EUC 2010, and a PC member of a lot of top conferences such as MobiCom, INFOCOM, and MOBIHOC. Yunhao and his research group are awarded best paper award or best paper candidate in many conferences such as PerCom 2007, MobiCom 2008, and ICPADS 2009. He won the grand prize of Hong Kong Best Innovation and Research Award in 2008. Being the vice-chair of ACM China Council, he is also an ACM Distinguished Speaker.

Zheng Yang received his bachelor of engineering degree in computer science at Tsinghua University in 2006 and a Ph.D. degree in computer science at Hong Kong University of Science and Technology (HKUST) in 2010. He is currently a member of Tsinghua National Lab for Information Science and Technology at Tsinghua University and a post-doc fellow at HKUST. His main research interests include wireless ad-hoc/sensor networks, pervasive computing, and internet of things. He has published more than 20 research papers in prestigious journals and conferences, including IEEE/ACM Transactions on Networking (ToN), IEEE Transactions on Parallel and Distributed Systems (TPDS), Journal of Computer Science and Technology (JCST), IEEE INFOCOM, IEEE ICDCS, IEEE RTSS, IEEE ICNP, ACM MobiCom, ACM SenSys. He is a member of the IEEE and the ACM.

Preface

With the popularity of wireless networks, location-based service (LBS) has quickly entered people's daily life. In practice, LBS has a large range of applications and often manifests in various forms in different types of networks. For example, E-911 in the USA, or corresponding E-112 in Europe, offers timely and accurate assistance to the emergency callers by locating them through mobile communication networks or global positioning system (GPS). Location information also plays a major role in modern asset management. Some companies, in particular hospitals, have deployed the WiFi-based solutions for real-time equipment locating and tracking, in order to increase equipment utilization and reduce overpurchasing costs. In addition, sensor network, a typical type of wireless ad hoc networks, has shown its great prospects of environmental monitoring, industrial sensing and diagnosis, battlefield surveillance, context-aware computing, and more. Autonomous localization of sensor nodes is essential since location makes the sensory data geographically meaningful. In all, many applications and services of wireless networks directly or indirectly rely on location information.

This book aims to provide a comprehensive and in-depth view of location-awareness technology in today's popular wireless networks. However, the obvious diversity of networks, from short-range bluetooth to long-range telecommunication network, makes it very challenging to organize materials. Although general principles exist, the implementation differs from network to network and application to application.

When composing the text, we have been thinking a lot what materials to include and how to organize them. Our thoughts come to the following two decisions. First, from the perspective of application, the book focuses on wireless ad hoc and sensor networks, in which the overwhelming majority of localization techniques are involved. Indeed, the techniques discussed in the book are quite versatile. Other types of networks, such as WLAN and 3G mobile network, are also mentioned. Second, to make it better understood, the book is basically organized around three step-by-step themes: location, localization, and localizability. Location-based applications are close to daily life and accordingly presented at the beginning.

Afterward, as the major part of the book, localization approaches are discussed in-depth. Other advanced topics, such as localizability and location privacy, are studied at last.

Book Organization

To begin with, the background of LBS and localization for wireless networks is presented in Chap. 1. Localization relies on the knowledge of physical world, in particular, the geometric relationship of network nodes. Chapter 2 discusses some popular ranging methods, including radio signal strength (RSS), time of arrival (ToA), time difference of arrival (TDoA), and hop counts. According to the physical measurements, one-hop positioning, as well as the related mathematical techniques of location computation, is presented in Chap. 3. Chapters 4 and 5 discuss the range-based and range-free localization approaches, respectively. Chapter 6 studies a key factor, error control, which determines the success of a localization approach in practice. Typically, location errors come from two sources: ranging noises and algorithm design, both of which are explained in detail. Chapter 7 presents the localization approaches for mobile networks, in which network nodes physically move and their locations change continuously. As we know, different approaches have different capabilities in terms of the number of nodes whose locations can be determined by a particular approach. Chapter 8 studies the issue of localizability that characterizes such capability in theory. With the development of LBS, location privacy is becoming crucial, which is discussed in Chap. 9.

The book discusses many up-to-date localization algorithms in considerable depth, yet makes their design and analysis accessible to all levels of readers. We emphasize the basic concepts and designs while keep the completeness. Each chapter presents a related topic and is independent of each other. When finishing the first two chapters, readers can select the remaining ones by their own interests. Each chapter ends with a summary or a comparative study, which provides a big picture and facilitates understanding.

Anticipated Audience

The book can serve as a guide book for the technicians and practitioners in the industry of real-time location systems (RTLS) and wireless networks. They can expect to obtain a comprehensive understanding of the field through reading the book, in order to compare and select localization solutions fulfilling various application requirements. Abundant references of the book open up a broader domain for advanced study. In addition, the book is tailored toward a textbook for college researchers and graduate students. For a one-semester gradate course, the main part includes three chapters (Chaps. 3–5) about one-hop positioning and multi-hop network localization. Chapter 6 can be used as a follow-on topic. When there is time, three independent chapters (Chaps. 7–9) can be added to course materials with freedom of choice. The book includes the state-of-the-art research results in many technical journals and conferences during its preparation. Readers can track trends and hot topics in the field.

Last but not least, the book purposefully accommodates the different backgrounds and career objectives of its reader. Specifically, it does not require a background of location-awareness technology. But as a technical book, we hope the readers have a basic knowledge of computer algorithms and networks.

Acknowledgments

During almost 10 years working on the topic of location system, we have benefited a great lot from our friends and colleagues. We wish to take this opportunity to thank them.

Many former and present graduate students have contributed greatly to the book. Some chapters are based entirely on their writing. We particularly thank Xiaoping Wang, Lirong Jian, and Junliang Liu. They have done great jobs.

We are grateful to many colleagues and peer researchers who shared with us exciting information. Especially, we thank Guohong Cao, Jiannong Cao, Guihai Chen, Tolga Eren, Jie Gao, David Goldenberg, Tian He, Jennifer Hou, Bill Jackson, Weijia Jia, Xiaohua Jia, Tibor Jordan, Wang-Chien Lee, Jianzhong Li, Qun Li, Xiang-Yang Li, Chenyang Lu, Guoqiang Mao, Abhishek Patil, Chunyi Peng, Yang Richard Yang, John Stankovic, Ivan Stojmenovic, Peng-Jun Wan, Jie Wu, Li Xiao, Dong Xuan, Baijian Yang, Yuanyuan Yang, Xiaodong Zhang, Tarek Abdelzaher, and Feng Zhao.

We thank all of those who have played in part in the preparation of this text. They include Prof. Lionel Ni, Prof. Renyi Xiao, Prof. Lei Chen, Dr. Mo Li, Dr. Jinsong Han, Dr. Yuan He, and Dr. Kebin Liu. The editor of this book, Brett Kurzman, of Springer provided many aspects of editorial support.

Finally, Yunhao Liu thanks his wife Jingyao Dai, and their parents Kai Liu, Yonghua Zhu, and Jianhua Zhang, for their unconditional support. Zheng Yang wishes to thank Chao Zhang, and his parents Aiguo Yang and Yajuan Gong, and his grandparents Jinggui Gong and Fenglan Niu, for their understanding and encouragement. We affectionately dedicate this book to them.

Yunhao Liu
Zheng Yang

Contents

Chapter 1
Introduction

"Location, Location, Location"

— anonymous

It's often said that the success of a retail store depends on three factors: location, location, and location. So do many services of wireless networks.

1.1 Location-Based Services

Location-based service (LBS) is a key-enabling technology of these applications and widely exists in today's wireless communication networks from the short-range Bluetooth to the long-range telecommunication networks, as illustrated in Fig. 1.1. In this book, we mainly study the location, localization, and localizability. In LBS, location information is generally considered to reveal the basic knowledge of the circumstance and background of service users. Localization is a process to compute the locations of wireless devices in a network. Localizability, answering whether or not the locations of nodes can be uniquely determined, plays a beneficial role on localization and a number of other network services. We focus on the three above-mentioned issues of wireless ad hoc and sensor networks, while the principles and techniques discussed in this book are of general purposes and can be used to other scenarios, such as WLAN, cellular networks, or other wireless networks.

1.1.1 Location-Based Applications

The proliferation of wireless and mobile devices has fostered the demand for context-aware applications, in which location is viewed as one of the most significant contexts. For example, pervasive medical care is designed to accurately record and manage patient movements [1, 2]; smart space enables the interaction between physical space and human activities [3, 4]; modern logistics has major concerns on

Y. Liu and Z. Yang, *Location, Localization, and Localizability: Location-awareness Technology for Wireless Networks*, DOI 10.1007/978-1-4419-7371-9_1,

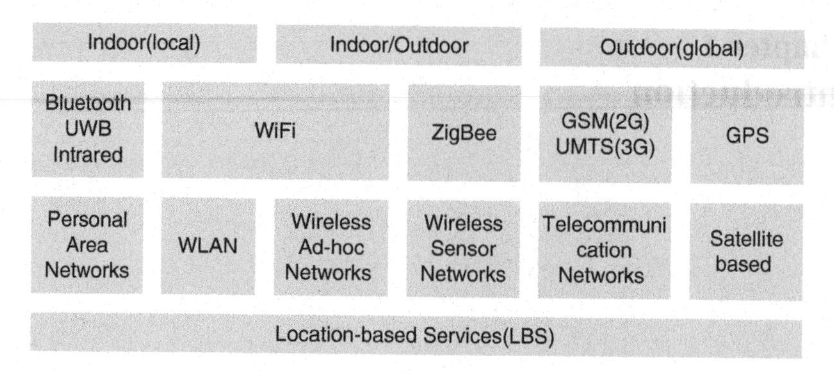

Fig. 1.1 Location-based services for a wide range of wireless networks

goods transportation, inventory, and warehousing [5, 6]; environmental monitoring networks sense air, water, and soil quality and detect the source of pollutants in real time [7–13]; and mobile peer-to-peer computing encourages content sharing and contributing among mobile hosts in the vicinity [14, 15]. A detailed survey on location-based applications can be found in [16, 17].

Recent technological advances have enabled the development of low-cost, low-power, and multifunctional sensor devices. These nodes are autonomous devices with integrated sensing, processing, and communication capabilities. With the rapid development of wireless sensor networks (WSNs), location information becomes critically essential and indispensable. The overwhelming reason is that WSNs are fundamentally intended to provide information on spatial–temporal characteristics of the physical world; hence, it is important to associate sensed data with locations, making data geographically meaningful. Almost all typical WSN applications, such as object tracking and environment monitoring, inherently rely on location information.

1.1.1.1 Motivating Example: GreenOrbs

We start our discussion from a forest surveillance system recently launched in China, in which location information is essentially important [10].

In the past decade, forest has received increasing attention from governments, scientists, industries, and people all over the world, due to its great significance in environmental protection, global climate change, and sustainable development. Forest is regarded as "the lung of the Earth," which is the major component on the Earth that absorbs carbon dioxide (CO_2) and release oxygen (O_2). Forest management and surveillance become important missions nowadays. Forestry applications usually require long-term, continuous, synchronized monitoring on huge measurement areas with diverse creatures and complex terrains.

On the other hand, wireless sensor networks (WSNs) in nature have some attractive features. WSNs can be designed to support large-scale deployments, continuous monitoring, and coordinated sensing at relatively low cost. More importantly, the sensor nodes can self-organize to accomplish their tasks without human supervision.

Under this circumstance, researchers have launched the GreenOrbs project [9, 10], a long-term, large-scale wireless sensor network system in the forest. The missions of GreenOrbs are twofold: On the one hand, GreenOrbs realizes all-year-round ecological surveillance in the forest, collecting various sensory data, such as temperature, humidity, illumination, and carbon dioxide titer. The collected information can be utilized to support various forestry applications. On the other hand, GreenOrbs is one of the latest efforts in the research community to build practical WSN systems. Through the real-world experience in GreenOrbs, researchers expect to explore the potential design space and scientific solutions of WSN.

The large-scale deployment of GreenOrbs system is carried out in the Tianmu Mountain, Zhejiang, China, as shown in Fig. 1.2. Researchers adopt TeloB motes with MSP430 processor and CC2420 radio. The sensor node software is developed based on TinyOS 2.1, using a globally synchronized duty-cycling mechanism for all the nodes. In each power-on period of the nodes' radios, researchers adopt the CTP protocol to collect the sensory data, whereas the beacon frequency is modified to save communication cost. Data disseminations from the sink are enabled to control the nodes' operational parameters, such as the transmission power, sampling frequency, duty ratio, and the length of a duty cycle. By April 2010, GreenOrbs has expanded to include 1,000+ nodes. The nodes using battery power will be kept in continuous operation for over 1 year.

Currently, GreenOrbs supports three typical applications: canopy closure estimate, fire risk evaluation, and forest microclimate observation. For all three applications, localization is important since the sensory data without locations are fairly meaningless. GPS is a straightforward solution but has two limitations for GreenOrbs uses. First, GPS signal is highly dynamic and unstable in forest environments, resulting in poor location accuracy. Second, it is costly to equip each sensor node a GPS receiver for such a large-scale system. In fact, GreenOrbs adopts a combined scheme that uses GPS and network localization simultaneously.

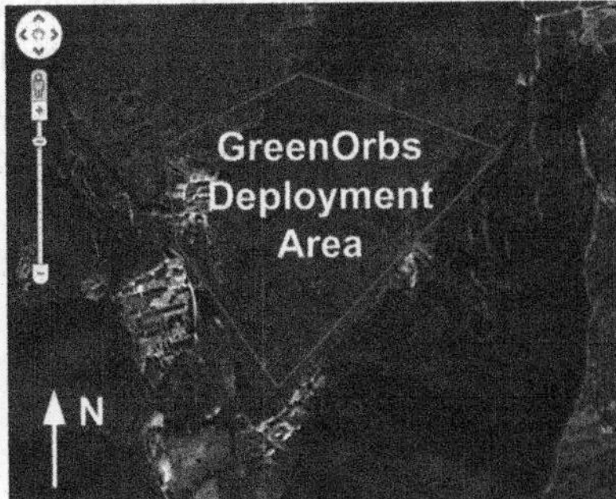

Fig. 1.2 GreenOrbs deployment

1.1.2 Location-Aided Network Functions

Location information also supports many fundamental network services, such as network routing, topology control, coverage, boundary detection, and clustering. A brief overview is as follows.

1.1.2.1 Routing

Routing is a process of selecting paths in a network along which to send data traffic. Many routing protocols for multihop wireless networks utilize physical locations to construct forwarding tables and send messages to the node closer to the destination in each hop [18]. Specifically, when a node receives a message, local forwarding decisions are made according to the positions of the destination and its neighboring nodes. Such geographic routing schemes require localized information, making the routing process stateless, scalable, and low-overhead in terms of route discovery.

1.1.2.2 Topology Control

Topology control is one of the most important techniques used in wireless ad hoc and sensor networks for saving energy and eliminating radio interference [19, 20]. By adjusting network parameters (e.g., the transmitting range), energy consumption and interference can be effectively reduced; meanwhile some global network properties (e.g., connectivity) can still be well retained. Certainly, using location information as a priori knowledge, geometry techniques (e.g., spanner subgraphs and Euclidean minimum spanning trees) can be immediately applied to topology control [19].

1.1.2.3 Coverage

Coverage reflects how well a sensor network observes the physical space; thus, it can be viewed as the quality of service (QoS) of the sensing function. Previous designs fall into two categories. The probabilistic approaches [21–23] analyze the node density for ensuring appropriate coverage statistically, but essentially have no guarantee on the result. In contrast, the geometric approaches [24] are able to obtain accurate and reliable results, in which location information is often necessary.

1.1.2.4 Boundary Detection

Boundary detection is to figure out the overall boundary of an area monitored by a WSN. There are two types of boundaries: the outer boundary showing the undersensed area, and the inner boundary indicating holes in a network deployment. The knowledge of boundary facilitates the design of routing, load balancing, and network management [25]. As a direct evidence, location information helps to identify border nodes and further depict the network boundary.

1.1.2.5 Clustering

To facilitate network management, researchers often propose to group sensor nodes into clusters and organize nodes hierarchically [26]. In general, ordinary nodes only talk to the nodes within the same cluster, and the intercluster communications rely on a special node in each cluster, which is often called cluster head. Cluster heads form a backbone of a network, based on which the network-wide connectivity is maintained. Clustering brings numerous advantages on network operations, such as improving network scalability, localizing the information exchange, stabilizing the network topology, and increasing the network life time. Among all possible solutions, location-based clustering approaches are greatly efficient by generating non-overlapped clusters. In addition, location information can also be used to rebuild clusters locally when new nodes join the network or some nodes suffer from hardware failure [26].

1.2 Introduction to Localization

One method to determine the location of a device is through manual configuration, which is often infeasible for large-scale deployments or mobile systems. As a popular system, global positioning system (GPS) is not suitable for indoor or underground environments and suffers from high hardware cost. Local positioning systems (LPs) rely on high-density base stations being deployed, an expensive burden for most resource-constrained wireless ad hoc networks.

Limitations of the existing positioning systems motivate a novel scheme of network localization, in which some special nodes (a.k.a. anchors or beacons) know their global locations and the rest determine their locations by measuring the geographic information of their local neighboring nodes. Such a localization scheme for wireless multihop networks is alternatively described as "cooperative," "ad hoc," "in-network localization," or "self localization," since network nodes cooperatively determine their locations by information sharing.

The terms of "known" and "unknown" nodes are referred to the nodes being aware and being aware of their locations, respectively. Suppose a specific positioning process in which an unknown node determines its location based on the information provided by a number of known nodes. The unknown node is also known as a target node or a to-be-located node, while the known nodes as reference nodes.

Almost all existing localization approaches basically consist of two stages: 1 measuring geographic information from the ground truth of network deployment and 2 computing node locations according to the measured data. Geographic information includes a variety of geometric relationships from coarse-grained neighbor-awareness to fine-grained internode rangings (e.g., distance or angle). Based on physical measurements, localization algorithms solve the problem that how the location information from beacon nodes spreads network-wide. Generally, the design of localization algorithms largely depends on a wide range of factors, including resource availability, accuracy requirements, and deployment restrictions, and no particular algorithm is an absolute favorite across the spectrum.

1.3 Book Organization

In Chap. 2, according to the capabilities of diverse hardware, we classify the measuring techniques into six categories (from fine grained to coarse grained): location, distance, angle, area, hop count, and neighborhood. Among them, the most powerful physical measurement is directly obtaining the position without any further computation. GPS is such a kind of infrastructure. We discuss the other five measurements in this chapter, with emphasis on the basic principles of measuring techniques. Basically, distance-related information can be obtained by radio signal strength or radio propagation time; angle information by antenna arrays; and area, hop count, and neighborhood information by the fact that radios only exist for nodes in vicinity.

Chapter 3 shows how to transform physical measurements to locations of nodes. Typically, this step takes place among a target node and its neighboring beacons. We name the process one-hop location estimation. Various kinds of optimization techniques are used for accurate location computation. In particular, we discuss the positioning methods for measurements of distance, time difference of arrival (TDoA), angle of arrival (AoA), and radio signal strength (RSS)-profiling. The distances from an unknown node to several references constrain the presence of this node, which is the basic idea of the so-called trilateration (or multilateration). TDoA measurement gives the difference of the time receiving the same signal on different reference nodes. Given a TDoA measurement Δt_{ij} and the coordinates of reference nodes i and j, they define one branch of a hyperbola whose foci are at the locations of reference nodes i and j. Hence, the unknown node must lie on the hyperbola. AoA measurement gives the bearing information of the two nodes. By combining the AoA estimates of two reference nodes, an estimate of the position can be obtained. RSS-profiling-based methods directly utilize RSS measurement data for location estimation. Since the RSS distribution of a set of anchor nodes is relatively stable over the spatial space, the RSS vector measured at an unknown node, defined as RSS finger print, reveals the physical location of the node.

Chapter 4 discusses the range-based localization approaches that adopt ranging techniques and use internode distance measurements to calculate the locations of nodes. According to the computation organization, we classify range-based localization algorithms into two main categories: centralized algorithms and distributed algorithms. Centralized algorithms take all distance measurements as the input and locate the entire wireless network in a single step. In contrast, in distributed algorithms, every node determines its location according to the information provided by its neighbors, to avoid the network-wide information exchange. This chapter reviews the state-of-the-art designs of range-based approaches. Multidimensional scaling (MDS), as a centralized algorithm, adopts statistical techniques and assigns locations of nodes such that the short-distance-apart nodes are close to each other in the localization results. We also present a well-known distributed solution, iterative trilateration (or sequential trilateration). It is computationally efficient and easy to implement, thus widely used in many systems. The variations of iterative trilateration, including bilateration and sweeps, are also discussed in Chap. 4.

Due to the hardware limitations, ranging is not always available for wireless devices. In such situations, range-free approaches are cost-effective alternatives, in which nodes merely know their neighbors (a.k.a. connectivity information). In Chap. 5, we show range-free approaches. Without direct distance ranging, the physical distance of a pair of nodes is estimated by the hop count or the proximity. The basic idea of hop count-based localization is to use hop by hop message delivery to calculate hop counts from nodes to anchors. The hop-count information is further converted to the distance estimates. Eventually, each node adopts trilateration or other methods to determine its location according to the estimated distances. Another possibility of range-free approaches is to explore the relative proximity of nodes. Although distance ranging is not available, the information that one node is closer to some node than others has great potentials to be used for localization. Both hop-count-and proximity-based solutions are discussed in Chap. 5.

Chapter 6 studies techniques that localization approaches use for error control. Although a number of ranging techniques are developed, noises and outliers are inevitable in distance measurements. Numerous simulations and experiments have suggested that the performances of many localization schemes would be drastically degraded if ranging errors are not handled properly. As a result, error control attracts a lot of research efforts. We first review the measurement accuracies of different ranging methods. Then, we discuss how noisy and outlier ranging results affect localization results in four aspects: uncertainty, nonconsistency, ambiguity, and error propagation. Finally, we present the state-of-the-art studies on error characterization, ambiguity elimination, location refinement, and outlier resistance, all of which have the goal of mitigating the negative impact of errors in distance measurements.

In Chap. 7, we concern the localization for mobile networks. Node mobility gives rise to new challenges of localization. The most straightforward and essential one is that localization is no longer a one-time task, but a continuous and repeated procedure. Many new localization algorithms have been proposed for mobile networks. We first introduce the Monte Carlo localization (MCL) algorithm, which casts the mobile localization problem as a Markov process. We then discuss the convex approximation localization (CAL) algorithms maintaining a convex polygon or circle to approximate the potential location of each node. We also discuss the moving-baseline localization (MBL) algorithm, whose emphasis is to construct a global view of the network from the perspective of each individual node. Finally, some techniques for universal localization (localizing static nodes and mobile nodes simultaneously) are depicted.

Chapter 8 presents the localizability issue of wireless networks. Network localizability is to determine whether or not a network is localizable given distance constrains. In recent years, this issue draws remarkable attentions from an increasing number of researchers. Based on rigidity theory, we analyze the reasons why the locations of some nodes in a network cannot be uniquely determined. In addition, we present two approaches for inductively constructing globally rigid graphs: trilateration and Wheel. Wheel is proved to be a nice substitute for trilateration, determining the locations of a larger number of nodes, more suitable for sparsely or moderately connected networks, and introducing no extra communication cost. At last, we discuss node localizability. Different from network localizability, node localizability focuses on whether or not a specific node is localizable given distance constrains. Actually, node localizability is a more general issue than network localizability. We investigate the state-of-the-art results on finding the condition for a node being localizable. In general, this is a new research area, and a number of open issues exist.

Location privacy is considered in Chap. 9. We begin with examining potential threats against location privacy. We then review major privacy protecting strategies, which fall into four categories: regulatory, privacy policies, anonymity, and obfuscation. Following a quick tour of these strategies, we dig into some anonymity-based approaches that use some tricks to fool the adversaries. Releasing location information anonymously (i.e., using a pseudonym instead of an actual identity) can prevent attackers from linking the location information to an actual identity. However, hiding the name is not enough. Certain regions of a space, such as desk location in an office, can be closely associated with certain identities, and hence can be used to deanonymize identities, although pseudonym is applied. In anonymity-based approaches, a trusted intermediary is introduced to coordinate users and to provide a large enough anonymity set, in which a certain identity cannot be distinguished from others. We present some typical approaches and make a comparison. At the end of Chap. 9, we provide several directions of future research on location privacy.

Chapter 2
Physical Measurements

It is absolutely infeasible to do localization without knowledge of the physical world. According to the capabilities of diverse hardware, we classify the measuring techniques into six categories (from fine grained to coarse grained): location, distance, angle, area, hop count, and neighborhood, as shown in Fig. 2.1.

Among them, the most powerful physical measurement is directly obtaining the position without any further computation. GPS is such a kind of infrastructure. Besides, the other five measurements are used in the scenarios of positioning an unknown node by giving some reference nodes. Distance and angle measurements are obtained by ranging techniques, while hop count and neighborhood are basically based on radio connectivity. In addition, area measurement relies on either ranging or connectivity depending on how the area constrains are formed.

2.1 Distance Measurements

Many physical quantities are distance related, such as the received radio signal strength, the propagation time of an acoustic signal. Investigating the physical characteristics of signals, researchers form the basic quantity-distance models that convert the measured signals to the physical distances. In this section, we mainly focus on the typical ranging techniques: radio signal strength (RSS), time of arrival (ToA), and time difference of arrival (TDoA).

2.1.1 Radio Signal Strength

RSS-based ranging techniques rely on the fact that the strength of radio signal diminishes during propagation. As a result, the understanding of radio attenuation helps to map signal strength to distance.

A common assumption is that the propagation distances d is much larger than the square of the antenna size divided by the wavelength. In an idealized free space, RSS is proved to be linear with the inverse square of the distance d between the

Y. Liu and Z. Yang, *Location, Localization, and Localizability: Location-awareness Technology for Wireless Networks*, DOI 10.1007/978-1-4419-7371-9_2,

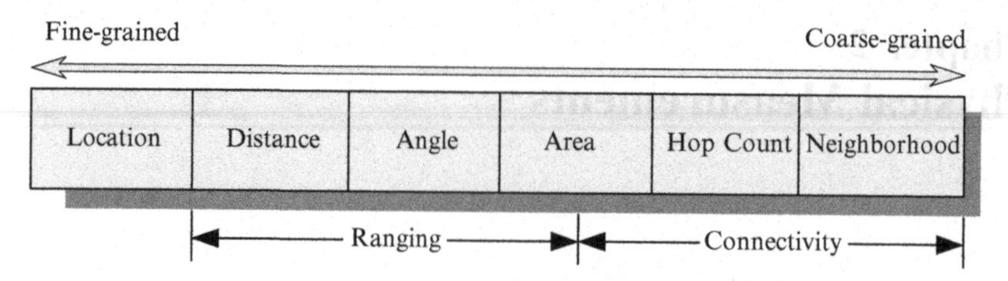

Fig. 2.1 Physical measurements

transmitter and the receiver. Let $P_r(d)$ denote the received power at distance d. The value of $P_r(d)$ follows the Friis equation [27]:

$$P_r(d) = \left(\frac{\lambda}{4\pi d}\right)^2 P_t G_t G_r,$$

where P_t is the transmitted power, G_t and G_r are the antenna gain of the transmitting and receiving antennas, respectively, and λx is the wavelength of the transmitter signal in meters.

In practice, several factors, such as shadowing and reflection, may affect the radio signal propagation as well as the received power. Unfortunately, these factors are environment dependent and unpredictable. As the shadowing effects cannot be precisely tracked, they are usually modeled as a log-normally distributed random variable. Considering the randomness, signal strength diminishes with distance according to a power law. One model used for wireless radios is as follows [28]:

$$P_r(d) = P_0(d_0) - \eta 10 \log_{10}\left(\frac{d}{d_0}\right) + X_\sigma,$$

where $P_r(d)$ denotes the received power at distance d and $P_0(d_0)$ denotes the received power at some reference distance d_0, η denotes the path-loss exponent, and X_σ denotes a log-normal random variable with variance σ^2 that accounts for fading effects. If the path-loss exponent for a given environment is known, the received signal strength can be used to estimate the distance. By this model, the maximum likelihood estimate of distance d is as follows [27]:

$$\hat{d} = d_0\left(\frac{P_r}{P_0(d_0)}\right)^{-1/\eta}.$$

In addition, the relationship between the estimated distance and the ground-truth distance is

$$\hat{d} = d10^{-\frac{X_\sigma}{10\eta}} = de^{-\frac{\alpha X_\sigma}{\eta}},$$

where $\alpha = \ln 10/10$. Hence, the expected value of the estimated distance is

$$E(\hat{d}) = \frac{1}{\sqrt{2\pi}\sigma}\int_{-\infty}^{\infty} d\, e^{-\alpha X_\sigma/\eta} e^{-X_\sigma/2\sigma^2} \mathrm{d}X_\sigma = d\, e^{(\alpha^2/2)(\sigma^2/\eta^2)}.$$

Thus the maximum likelihood estimate is biased from the ground-truth distance. Hence, an unbiased estimate is given by

$$\hat{d} = d_0\left(\frac{P_r}{P_0(d_0)}\right)^{-1/\eta} \mathrm{e}^{-(\alpha^2/2)(\sigma^2/\eta^2)}.$$

The ranging noise occurs because radio propagation tends to be highly dynamic in complicated environments. Although RSS-based ranging contains noises on the order of several meters (or even worse performance) [29], it is widely used in many real-world systems because RSS is a relatively "cheap" solution without any special hardware, as all nodes are supposed to have radios. It is believed that more careful physical analysis of radio propagation may allow better use of RSS data. Nevertheless, the breakthrough technology is not there today.

2.1.2 *Time of Arrival (ToA)*

For a signal with known velocity (e.g., acoustic signal), measuring the propagation-induced time can straightforwardly indicate the transmitter–receiver separation distance. The key issue of this mechanism is to accurately measure the time of arrival (ToA). There are two categories of ToA-based distance measurement: the one-way propagation time estimation and the round-trip propagation time estimation.

1) One-way propagation time estimation
Propagation delay, which can be calculated as $t_i - t_0$, is the time lag between the departure of a signal from a transmitter and the arrival at a receiver; in other words, it is the amount of time required for a signal to travel from a transmitter to a receiver. Assuming the speed of a signal v, the transmitter–receiver distance can be calculated by $d = v\,(t_i - t_0)$.

In the basic scheme of ToA, the receiver needs to know the time when the signal is sent from the transmitter. One method to release the requirement of time synchronization is the combined use of signals with different speeds, such as the ultrasound/acoustic and radio signals [30–32].

In such a scheme, each node is equipped with a speaker and a microphone, as illustrated in Fig. 2.2. Some systems use ultrasound while others use audible frequencies. The general ranging technique, however, is independent of any particular hardware.

The idea of ToA ranging is conceptually simple, as illustrated in Fig. 2.3. The transmitter first emits a radio signal. It waits some fixed internal of time, t_{delay}

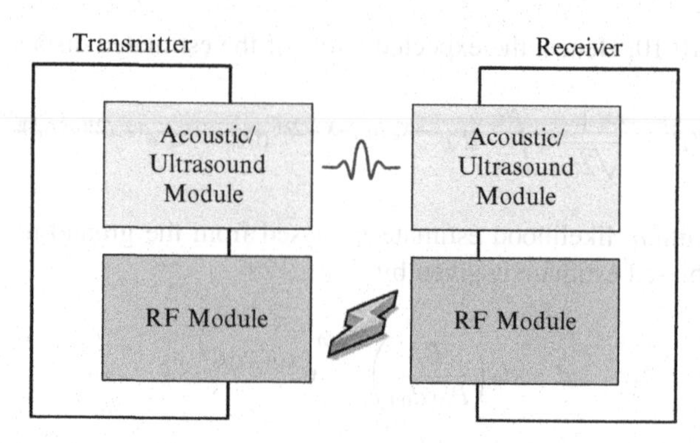

Fig. 2.2 ToA hardware model

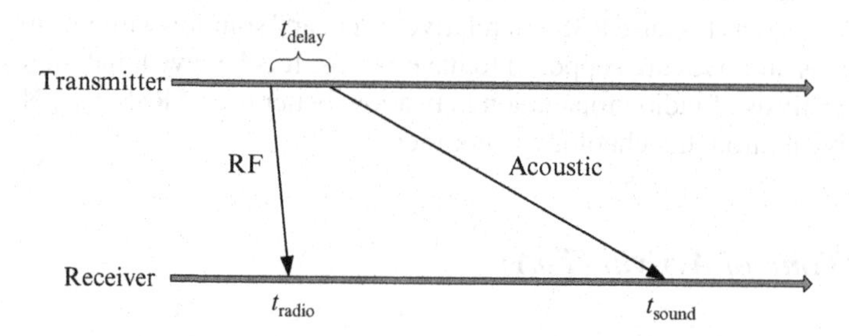

Fig. 2.3 ToA computation model

(which might be zero), and then produces a fixed pattern of "chirps" on its speaker. When the receivers hear the radio signal, they record the current time, t_{radio}, and turn on their microphones. When their microphones detect the chirp pattern, they again record the current time, t_{sound}. Once they have t_{radio}, t_{sound}, and t_{delay}, the receivers can compute the transmitter–receiver distance d by

$$d = \frac{v_{radio} v_{sound}}{v_{radio} - v_{sound}} (t_{sound} - t_{radio} - t_{delay}),$$

where v_{radio} and v_{sound} denote the speed of radio and sound waves, respectively. Since radio waves travel substantially faster than sound in air, the distance is then estimated as $d = v_{sound}\,(t_{sound} - t_{radio} - t_{delay})$. If radio and acoustic signals are designed to be transmitted simultaneously (i.e., $t_{delay} = 0$), the estimation can be further simplified as $v_{sound}\,(t_{radio} - t_{sound})$.

ToA methods are impressively accurate under line-of-sight conditions. For instance, it is claimed in [31] that distance can be estimated with error no more than a few centimeters for node separations under 3 m. The cricket ultrasound system [30] can obtain centimeter accuracy over about 10-m range in indoor environments.

2) Round-trip propagation time estimation
One-way propagation time estimation requires synchronization between the nodes, because the computation relies on the timestamps recorded in both the nodes. One way to avoid synchronization is to use round trip time (RTT). In RTT measurement, nodes only need to report local time duration instead of the timestamps. The RTT estimate between two nodes, labeled as A and B, is as follows. Node A transmits a packet to node B. After receiving this packet, node B delays t_{delay}, and then replies node A by sending an acknowledgment packet. The RTT at A is determined by $t_{RT} = 2t_{\text{flight}} + t_{\text{delay}}$, where t_{flight} denotes the distance-induced propagation time of the signal. When node B reports the measured delay t_{delay}, node A can compute the time of signal propagation by $t_{\text{flight}} = (t_{\text{RT}} - t_{\text{delay}})/2$. However, RTT measurement suffers from the clock drift between the nodes, especially when t_{flight} is of the same level of the resolution of t_{RT} and t_{delay} measurements.

2.1.2.1 Symmetric Double Sided Two-Way Ranging (SDS-TWR)

When we adopt radio signal for the ToA-based distance measurement, the ranging mainly relies on the resolution of time measurement. There are two main sources of errors: the multipath effect and the time synchronization. Nanotron technologies proposes SDS-TWR to address such issues [33].

SDS-TWR adopts chirp spread spectrum (CSS) to provide fine resolution of a few nanoseconds for signal detection in spite of the multipath propagation and noises. CSS is a customized application of multidimensional multiple access (MDMA) for the requirements of battery-powered applications, where the reliability of the transmission and low power consumption are of special importance. CSS operates in the 2.45 GHz ISM band and achieves a maximum data rate of 2 Mbps. Each symbol is transmitted with a chirp pulse that has a bandwidth of 80 MHz and a fixed duration of 1 μs.

To avoid time synchronization, the elapsed time is measured by round trip time (RTT). RTT is the time duration between the timestamp of sending a ranging signal and that of the acknowledgement. RTT uses highly predictable hardware-generated acknowledgement packets where MAC processing time assumed to be equal on both nodes. Note that the timestamps are processed on the physical layer, not on the application layer. Figure 2.4 illustrates the measurement procedure, where we show

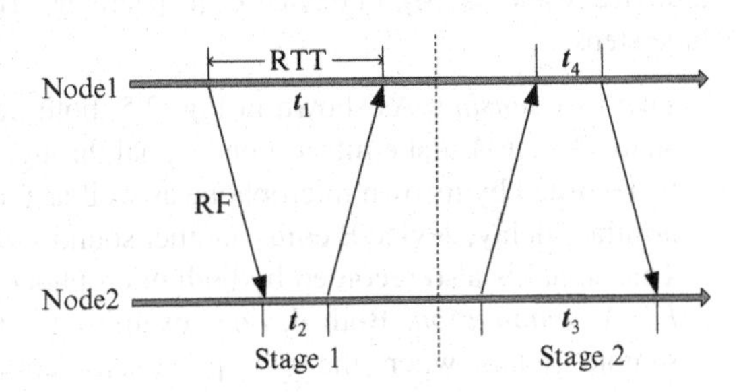

Fig. 2.4 Symmetric double sided two-way ranging

the packet transfer according to the time lines of two nodes. There are two stages for SDS-TWR, each of which is a RTT measurement.

In stage 1, time measurement of node 2 begins only when it receives a packet from node 1 and then stops when it sends a packet back to node 1. Thus, the distance between node 1 and node 2 is given by

$$d = v(t_1 - t_2)/2,$$

where v denotes the speed of radio signal. Nevertheless, such a scheme suffers clock drift between the two nodes, because off-the-shelf oscillators can only provide timing resolution of several nanoseconds. To mitigate the clock drift, SDS-TWR conducts the ranging measurement twice and symmetrically. As shown in Fig. 2.4, the first ranging measurement is calculated based on a round trip from node 1 to node 2 and back to node 1. The second measurement is calculated based on a round trip from node 2 to node 1 and back to node 2. This double-sided ranging measurement zeroes out the errors of the first order due to the clock drift. Hence, the distance estimate is given by

$$d = v[(t_1 - t_2) + (t_3 - t_4)]/4.$$

2.1.2.2 BeepBeep

Recently, researchers [34] observe that two intrinsic uncertainties in ToA can contribute to ranging inaccuracy: the possible misalignment between the sender timestamp and the actual signal emission, and the possible delay of a sound signal arrival being recognized at the receiver. To eliminate such uncertainties, round-trip measurement techniques are introduced.

In general, many factors can cause uncertainties in a real system, such as the lack of real-time control, software delay, interrupt handling delay, and system loads. These factors, if not controlled, can easily add up to several milliseconds on average, which translates to several feet of ranging error.

We show the general system model of BeepBeep design [34] in Fig. 2.5, in which each device is equipped with a speaker and a microphone, denoted by S_A, M_A in device A and S_B, M_B in device B, respectively. BeepBeep ranging scheme takes three steps:

1. *Two-way sensing*. As shown in Fig. 2.5, both devices are initially in recording state. Device A first emits a sound signal through its speaker S_A. This signal will be recorded by its own microphone as well as the other device B. Then, after an arbitrary delay, device B emits another sound signal back through its speaker S_B. This signal is also recorded by both microphones on the two devices.
2. *EToA computation*. Both devices examine their recorded data and locate the sample points when the two previously emitted signals arrive. The time

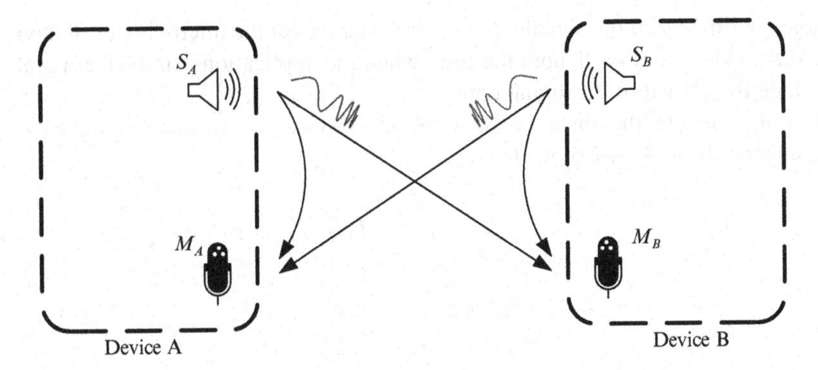

Fig. 2.5 The system model of the BeepBeep design

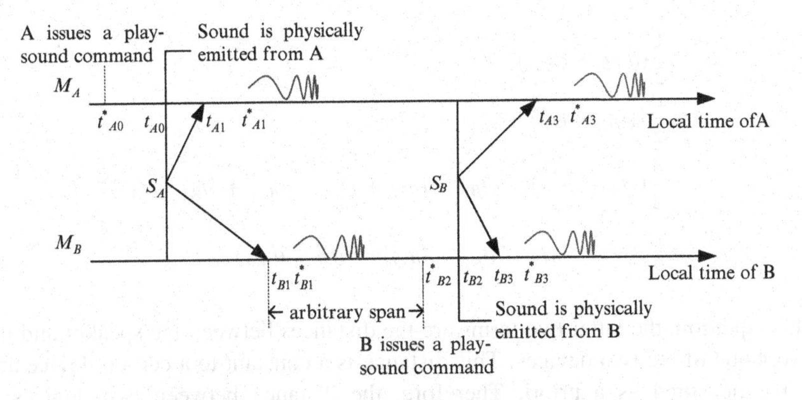

Fig. 2.6 Illustration of event sequences in BeepBeep ranging procedure

difference between these two signals is denoted as elapsed time between the two time of arrivals (EToA). When the EToA is computed, the two devices will exchange their locally measured EToA.

3. *Distance estimation*. The distance between the two devices is computed based on these two values of EToA.

Figure 2.6 shows the signal transmission procedure and timing relation among events in the first stage. Two time lines are drawn in the figure with the upper one representing the local time of device *A* and the bottom one the local time of device *B*. Let t^*_{A0} denote the time when device *A* instructs its speaker to emit the sound signal. Due to the sending uncertainty, however, the actual time when the speaker physically emits might be t_{A0}. The time the signal arrives at the microphones of devices *A* and *B* is marked t_{A1} and t_{B1}, respectively. Again, due to the receiving uncertainty, applications on device *A* and *B* may obtain these signal data only at time t^*_{A1} and t^*_{B1}. Similarly, let t^*_{B2} and t_{B2} denote the time when device *B* instructs to send out a sound signal and when the signal is physically out; t_{A3} and t_{B3}

denote the time when the signal from device B arrives at the microphones of device A and B; and t^*_{A3} and t^*_{B3} denote the time when the applications on device A and B conclude the arrival of the signal data.

Let $d_{x,y}$ denote the distance between the device x's speaker to device y's microphone. From Fig. 2.6, we have

$$\begin{aligned} d_{A,A} &= c(t_{A1} - t_{A0}), \\ d_{A,B} &= c(t_{B1} - t_{A0}), \\ d_{B,A} &= c(t_{A3} - t_{B2}), \\ d_{B,B} &= c(t_{B3} - t_{B2}), \end{aligned}$$

where c is the speed of sound. Then, the distance D between the two devices can be approximated as

$$\begin{aligned} D &= \frac{1}{2}(d_{A,B} + d_{B,A}) \\ &= \frac{c}{2}((t_{B1} - t_{A0}) + (t_{A3} - t_{B2})) \\ &= \frac{c}{2}((t_{A3} - t_{A1}) - (t_{B3} - t_{B1}) + (t_{B3} - t_{B2}) + (t_{A1} - t_{A0})) \\ &= \frac{c}{2}((t_{A3} - t_{A1}) - (t_{B3} - t_{B1}) + d_{B,B} + d_{A,A}). \end{aligned}$$

In this equation, the latter two terms are the distances between the speaker and the microphone of the two devices. This distance is a constant to a certain device and can be measured as a priori. Therefore, the distance between two devices is determined solely by the first two terms, which are actually the EToA values measured on device A and B, respectively. Note that EToA is calculated by each individual device independently, i.e., without referring any timing information on the other device, so that no clock synchronization between devices is needed. Moreover, due to the self-recording strategy, all time measurements are associated with the arrival instants of the sound signals, and, therefore, the sending uncertainty is also removed.

Obtaining the exact time instance when the signal arrives is difficult due to the indeterministic latency introduced by hardware and software (receiving uncertainty). Hence, the values of t_{A0}, t_{A1}, t_{A3}, t_{B1}, t_{B2}, and t_{B3} cannot be accurately measured. BeepBeep solves this issue by not referring to any local clock while inferring timing information directly from recorded sound samples.

As the received sound signal is always sampled at a fixed frequency (represented by f_s) by the A/D converter, BeepBeep directly obtains EToA by counting the sample number between the two ToAs of signals from recorded data, without dealing with the local clock of the end system. Thus, the accuracy depends on the fidelity of the recording module. Since all the sound signals are recorded, BeepBeep only needs to check the recorded data and identify the first sample point of each signal. Then, EToA is obtained by counting the number of samples between the two sound signals.

With sample counting, the above equation can be rewritten as

$$D = \frac{c}{2}\left(\frac{n_{A3} - n_{A1}}{f_{sA}} - \frac{n_{B3} - n_{B1}}{f_{sB}}\right) + K,$$

where n_x denotes the index of the sample point at instant t_x, f_{sA} and f_{sB} are the sampling frequency of device A and B, respectively, and $K = d_{B,B}+d_{A,A}$ is a constant. Assume the sampling frequency to be 44.1 kHz, since the 44.1 kHz sampling frequency is the basic, de facto standard that almost every sound card supports. In this case, we have $f_{sA} = f_{sB}$, and the above equation can be simplified to

$$D = \frac{c}{2f_s}((n_{A3} - n_{A1}) - (n_{B3} - n_{B1})) + K.$$

From this equation, the measurement granularity is positively proportional to the sound speed c and inversely proportional to the sampling frequency f_s. Take a typical setting of $c = 340$ m/s and $f_s = 44.1$ kHz, the distance granularity is then about 0.77 cm.

The distance granularity shows the best accuracy for BeepBeep system. In practice, due to several constraints, such as the signal to noise ratio, the multipath effects, and signal distortion, BeepBeep can achieve 1 or 2 cm accuracy. Experiments show that the operational range for the indoor cases is around 4 m and that for outdoor cases is in general larger than 10 m.

Being accurate, ToA systems are generally constrained by the line-of-sight condition, which is often difficult to meet in some environments. In addition, ToA systems perform better when they are calibrated properly, since speakers and microphones never have identical transmission and reception characteristics. Furthermore, the speed of sound in air varies with air temperature and humidity, which introduce inaccuracy into distance estimation. Acoustic signals also show multipath propagation effects that may impact the accuracy of signal detection. These can be mitigated to a large extent using simple spread-spectrum techniques [35]. The basic idea is to send a pseudorandom noise sequence as the acoustic signal and use a matched filter for detection, instead of using a simple chirp and threshold detection.

By designing BeepBeep, a high-accuracy acoustic-based ranging system, the localization can achieve 1 or 2 cm accuracy within a range of more than 10 m, which is so far the best result of ranging with off-the-shelf devices. Many localization algorithms use ToA simply because it is dramatically more accurate than radio-only methods. The trade-off is that nodes must be equipped with acoustic transceivers in addition to radio transceivers.

2.1.3 Time Difference of Arrival (TDoA)

When multiple reference nodes are available, there is a category of measurements called TDoA. The transmitter sends a signal to a number of receivers at known locations. Then, the receivers record the arrival time of the signal, as illustrated in Fig. 2.7. The location of the transmitter is computed by the difference of the recorded arrival timestamps. The TDoA between a pair of receivers i and j is given by

$$\Delta t_{ij} \triangleq (t_i - t_0) - (t_j - t_0) = t_i - t_j = \frac{1}{c}(||r_i - r_t|| - ||r_j - r_t||) \quad i \neq j$$

where t_0 is the time when the signal is sent from the transmitter (locating at r_t), t_i and t_j are the times when the signal is received at receivers i (locating at r_i) and j (locating at r_j), respectively, c is the speed of the signal, and $||\cdot||$ denotes the Euclidean norm. TDoA is also known as range difference since the speed of signal is assumed to be known as a priori.

Accurate TDoA measurement relies on two issues, time synchronization of receivers and signal detection, both of which are well known and still challenging. In the TDoA scheme, receivers need to be precisely synchronized to make the time difference $(t_i - t_j)$ valid. Even tiny errors of synchronization can totally destroy the final location results since the commonly used wireless signals travel fast (e.g., about 343 m/s for acoustic signals) or ultimately fast (e.g., $3 \times 10^8 \times$m/s for radio signals).

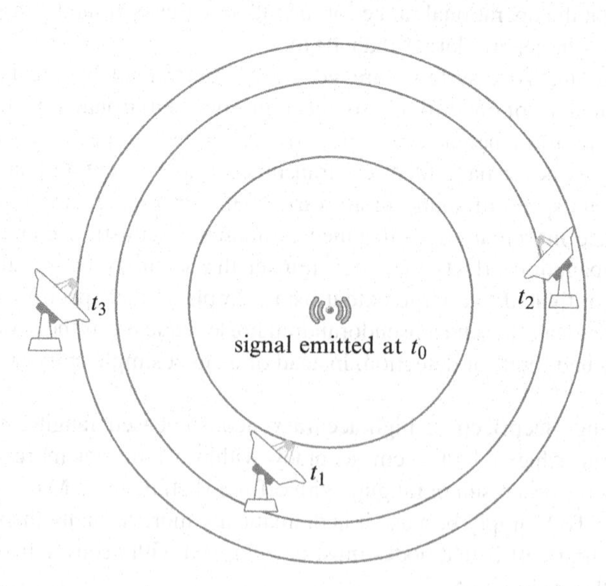

Fig. 2.7 TDOA measurement

Measuring the TDoA of a signal at two receivers at separate locations is a relatively mature field. The most widely used method is the generalized cross-correlation method [36], where the cross-correlation function between two signals s_i and s_j received at receivers i and j is given by integrating the lag product of two received signals for a sufficiently long time period T:

$$\rho_{i,j}(\tau) = \frac{1}{T}\int_0^T s_i(t)s_j(t-\tau)\mathrm{d}t.$$

TDoA only requires the receivers to be synchronized and does not demand any synchronization between the transmitter and the receivers. However, intensive computation is introduced and performed at receivers. Hence, it especially suits for the networks with powerful infrastructures, such as the cellular network.

2.2 Angle Measurement

Another possibility for localization is the use of angular estimates instead of distance estimates. In trigonometry and geometry, triangulation is the process of determining the location of a point by measuring angles to it from two known reference points (as illustrated in Fig. 2.8), using the law of sines. Triangulation is once used to find the coordinates and sometimes the distance from a ship to the shore.

The angle of arrival (AoA), a.k.a., direction of arrival (DOA), measurement is typically gathered using radio or microphone arrays, which allow a receiver to determine the direction of a transmitter. Suppose, several (3–4) spatially separated microphones hear a single transmitted signal. By analyzing the phase or time difference between the signal's arrival's at different microphones, it is possible to discover the AoA of the signal.

These methods can obtain accuracy within a few degrees [37]. A very simple localization technique, involving three rotating reference beacons at the boundary of a sensor network providing localization for all interior nodes, is described in [38].

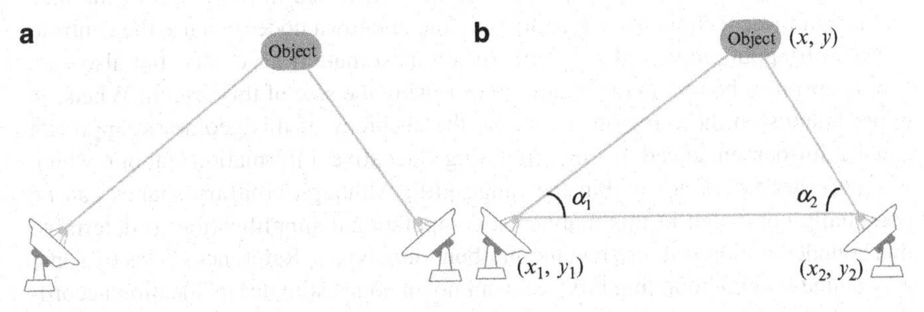

Fig. 2.8 Angle measurement

Unfortunately, AoA hardware tends to be bulkier and more expensive than ToA or TDoA ranging hardware, since each node must have one speaker and several microphones. Furthermore, the need of spatial separation between microphones is difficult to be accommodated in small size devices.

2.3 Area Measurement

If the radio or other signal coverage region can be described by a geometric shape, this can be used to provide location estimates by determining which geometric areas that a node is in. The basic idea of area estimation is to compute the intersection of all overlapping coverage regions and choose the centroid as the location estimate. Along with the increasing number of constraining areas, higher localization accuracy can be achieved.

According to how the area is estimated, we classify the existing ideas into two categories: single reference area estimation and multireference area estimation.

2.3.1 *Single Reference Area Estimation*

Single reference estimation means that areas are obtained in a pairwise manner, i.e., the information of a geometric area comes from only one reference at each stage. For instance, the region of radio coverage may be upper bounded by a circle of radius $R_{\max}$. In other words, if node B hears node A, it knows that it must be no more than a distance $R_{\max}$ from A. If an unknown node hears from several reference nodes, it can determine that it must lie in the geometric region described by the intersection of circles of radius $R_{\max}$ centered at reference nodes, as illustrated in Fig. 2.9a. This can be extended to other scenarios. For instance when both the lower bound $R_{\min}$ and the upper bound $R_{\max}$ can be determined, based on the received signal strength, the shape of a single node's coverage is an annulus, as illustrated in Fig. 2.9c; when an angular sector $(\theta_{\min}, \theta_{\max})$ and a maximum range $R_{\max}$ can be determined, the shape for a single node's coverage would be a cone with given angle and radius, as illustrated in Fig. 2.9d.

Localization using geometric regions is first described in [39]. One of the nice features of these techniques is that not only the unknown nodes can use the centroid of the overlapping region as a specific location estimate if necessary, but also they can determine a bound on the location error using the size of this region. When the upper bounds on these regions are tight, the accuracy of this geometric approach can be further enhanced by incorporating "negative information" about which reference nodes are not within the range [40]. Although arbitrary shapes can be potentially computed in this manner, a computational simplification to determine this bounded region is to use rectangular bounding boxes. Reference nodes by some way define several bounding boxes; an unknown node estimates its location according to the intersection of all boxes, which can be efficiently computed.

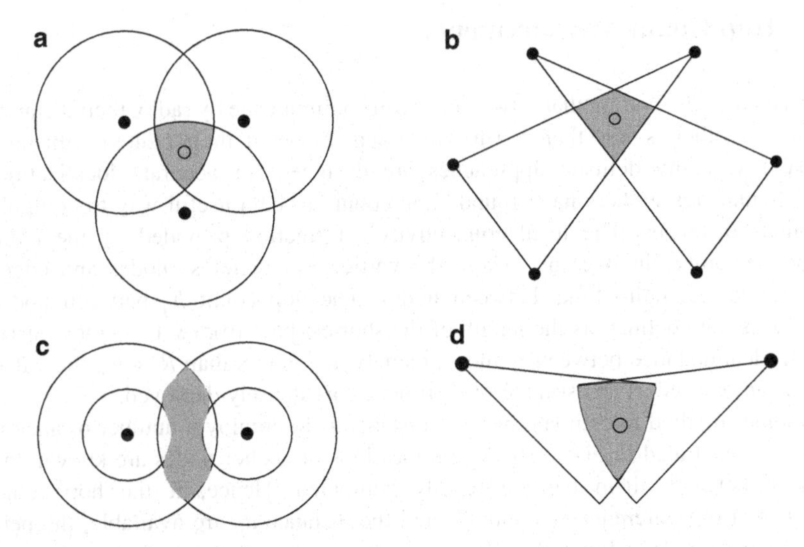

Fig. 2.9 Area measurements

2.3.2 *Multireference Area Estimation*

Another approach of area estimation is the approximate point in triangle (APIT) technique [41]. Its novelty lies in that regions are defined as triangles between different sets of three reference nodes, rather than the coverage of a single node.

APIT consists of two key processes: triangle intersection and point in triangle (PIT) test. Nodes are assumed to hear a fairly large number of beacons. A node forms some number of "reference triangles": the triangle formed by three arbitrary references. The node then decides whether it is inside or outside a given triangle by PIT test. Once the process is complete, the node finds the intersection of the reference triangles that contain it and chooses the centroid as its position estimate, as illustrated in Fig. 2.9b. During process, APIT does not assume that nodes can range to these beacons.

The PIT test is based on geometry. For a given triangle with points A, B, and C, a point M is outside triangle ABC, if there exists a direction such that a point adjacent to M is further/closer to points A, B, and C simultaneously. Otherwise, M is inside triangle ABC. Unfortunately, given that typically nodes cannot move, an approximate PIT test is proposed based on two assumptions. The first one is that the range measurements are monotonic and calibrated to be comparable but are not required to produce distance estimates. The second one assumes sufficient node density for approximating node movement. If no neighbor of M is further from/closer to all three anchors A, B, and C simultaneously, M assumes that it is inside triangle ABC. Otherwise, M assumes it resides outside this triangle. In practice, however, this approximation does not realize the PIT test well. Nevertheless, APIT provides a novel point of view to do localization based on area estimation.

2.4 Hop Count Measurements

Based on the observation that if two nodes can communicate by radio, their distance from each other is less than R (the maximum range of their radios) with high probability, many delicate approaches are designed for accurate localization. In particular, researchers have found "hop count" to be a useful way to compute internode distances. The local connectivity information provided by the radio defines an unweighted graph, where the vertices are wireless nodes and edges represent direct radio links between nodes. The hop count h_{ij} between nodes s_i and s_j is then defined as the length of the shortest path from s_i to s_j. Obviously, the physical distance between s_i and s_j, namely, d_{ij}, is less than $R \times h_{ij}$, the value which can be used as an estimate of d_{ij} if nodes are densely deployed.

Another method to estimate per-hop distance is to employ a number of anchor nodes, as illustrated in Fig. 2.10. As the locations of anchor nodes are known, the distance between them can be readily computed. Hence, if the hop count h_{ij} between two references (s_i and s_j) and the distance d_{ij} are available, the per-hop distance can be estimated as $d_{\text{hop}} = d_{ij}/h_{ij}$.

Due to the hardware limitations and energy constraints of wireless devices, hop-count-based localization approaches are cost-effective alternatives to range-based approaches. Since there is no way to measure physical distances between nodes, existing hop-count-based approaches largely depend on connectivity measurements with a high density of anchors.

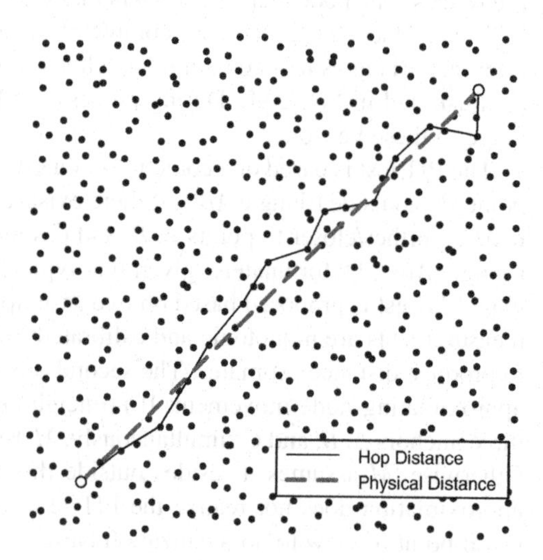

Fig. 2.10 Hop count measurement

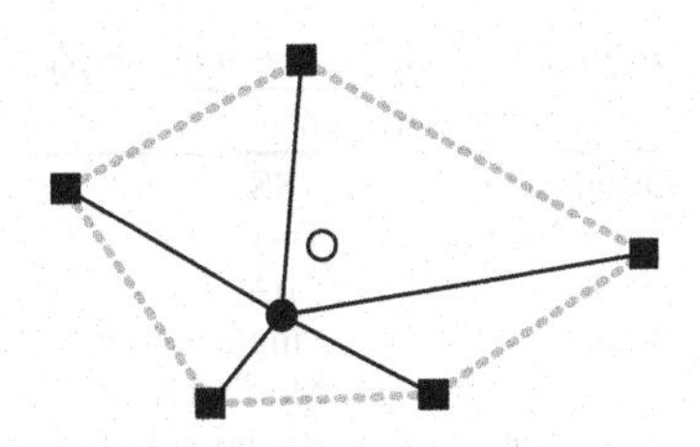

Fig. 2.11 *k*-neighbor proximity

2.5 Neighborhood Measurement

Radio connectivity can be considered economic since no extra hardware is required. Perhaps the most basic location technique is that of one-neighbor proximity, involving a simple decision of whether two nodes are within reception range of each other. A set of reference nodes is placed in the network with some nonoverlapping (or nearly nonoverlapping) subregions. Reference nodes periodically emit beacons including their location IDs. An unknown node uses the received location information as its own location, achieving a course-grained localization. The major advantage of this single-neighbor proximity approach is the simplicity of computation.

The neighborhood information can be more useful when the density of reference nodes is sufficiently high that there are several reference nodes within the range of an unknown node. Let there be k reference nodes within the proximity of the unknown node, as illustrated in Fig. 2.11. Suppose black squares are references and the black circle is the real location of the unknown node. We use the centroid (denoted by the hollow circle) of the polygon constructed by the k reference nodes as the estimated position of the unknown node. This is actually a k-nearest-neighbor approximation in which all reference nodes have equal weights.

This simple centroid technique has been investigated using a model with each node having a simple circular range R in an infinite square mesh of reference nodes spaced a distance d apart [42]. It is shown through simulation that, as the overlap ratio R/d is increased from 1 to 4, the average error in localization decreases from $0.5d$ to $0.25d$.

The k-neighbor proximity approach inherits the merit of computational simplicity from the single-neighbor proximity approach; while at the same time, it provides more accurate localization results than the single-neighbor proximity statistically.

2.6 Summary

In this section, a comparative study is presented for the existing physical measurement approaches. Table 2.1 provides an overview of these approaches in terms of accuracy, hardware cost, and environment requirements. All approaches have their own merits and drawbacks, making them suitable for different applications.

Table 2.1 Comparative study of physical measurements

Physical Measurements		Accuracy	Hardware cost	Computation cost
Distance	RSS	Median	Low	Low
	ToA	High	High	Low
Angle	AoA	High	High	Low
Area	Single reference	Median[a]	Median[a]	Median
	Multireference	Median[a]	Median[a]	High
Hop count	Per-hop distance	Median	Low	Median
Neighborhood	Single neighbor	Low	Low	Low
	Multineighbor	Low	Low	Low

[a]Depends on the diverse geometric constrains

Recent technical advances foster a novel ranging approaches. Ultra-wideband (UWB) is a radio technology that can be used at very low energy levels for short-range high-bandwidth communications by using a large portion of the radio spectrum [43]. It has relative bandwidth larger than 20% or absolute bandwidth of more than 500 MHz. Such wide bandwidth offers a wealth of advantages for both communications and ranging applications. In particular, a large absolute bandwidth offers high resolution with improved ranging accuracy of centimeter level.

UWB has a combination of attractive properties for in-building location systems. First, it is a non-line-of-sight technology with a range of a few tens of meters, which makes it practical to cover large indoor areas; second, it is easy to filter the signal to minimize the multipath distortions that are the main cause of inaccuracy in RF-based location systems. With conventional RF, reflections in in-building environments distort the direct path signal, making accurate pulse timing difficult; while with UWB, the direct path signal can be distinguished from the reflections. These properties provide a good cost-to-performance ratio of all available indoor location technologies.

In some positioning systems, two or more types of physical measurements, studied in previous subsections, are used simultaneously in order to obtain more information about the target node and increase the accuracy and robustness of positioning. Examples of such multimodal (or hybrid) scheme include ToA/AoA [44], ToA/RSS [45], TDoA/AoA [46], and ToA/TDoA [47]. Recently, some progresses from computational geometry reveal the great potential of multimodal measurements, regarding localization accuracy. With the rapid development of integrated circuits, multimodal measurement has been available on many wireless devices, especially sensor motes.

In all ranging algorithms discussed above, nodes should actively participate in the ranging process, i.e., sending or receiving radio signals, or measuring physical data. For some applications, however, the to-be-located objects cannot join the process, and it is also difficult to attach networked nodes to them. One typical application is intrusion detection, in which it is impossible and unreasonable to equip intruders with locating devices. To tackle this issue, recently a novel concept of device-free localization, also called transceiver-free localization, is proposed [48, 49].

Device-free localization is envisioned to be able to detect, localize, track, and identify entities free of devices and works by processing the environment changes collected at scattering monitoring points. Existing work focuses on analyzing RSS changes, and often suffers from high false positives. How to design a device-free localization system which can provide accurate locations is a challenging and promising research problem.

Chapter 3
One-Hop Location Estimation

This chapter discusses how to transform physical measurements to locations of nodes. This step is a basic and essential building block of all localization approaches. Typically, it takes place among a target node and its neighboring beacons. Thus, we name it one-hop location estimation. Various kinds of optimization techniques are used in this step for accuracy.

In particular, we discuss the positioning methods for measurements of distance, TDoA, AoA, and RSS-profiling. The distances from an unknown node to several references constrain the presence of this node, which is the basic idea of the so-called multilateration. TDoA measurement gives the difference of the time receiving the same signal on different reference nodes. Given a TDoA measurement Δt_{ij} and the coordinates of reference nodes i and j, they define one branch of a hyperbola whose foci are at the locations of reference nodes i and j. Hence, the unknown node must lie on the hyperbola. AoA measurement gives the bearing information of the two nodes. By combining the AoA estimates of two reference nodes, an estimate of the position can be obtained. RSS-profiling-based methods directly utilize RSS measurement data for location estimation. Since the RSS distribution of a set of anchor nodes is relatively stable over the spatial space, the RSS vector measured at an unknown node, defined as RSS finger print, reveals the physical location of the node.

3.1 Distance-Based Positioning Techniques

Multilateration is the process of locating an object according to distance measurements. Note that the word "multilateration" has different meanings in the context of localization, and in this book it refers to the distance-based positioning technique. Figure 3.1 shows an example of trilateration, a special form of multilateration which utilizes exact three references. The object to be localized (the soft dots) measures the distances from itself to three references (the solid squares). Obviously, the object should locate at the intersection of three circles centered at each reference position. The result of trilateration is unique as long as three references are nonlinear.

Y. Liu and Z. Yang, *Location, Localization, and Localizability: Location-awareness Technology for Wireless Networks*, DOI 10.1007/978-1-4419-7371-9_3,

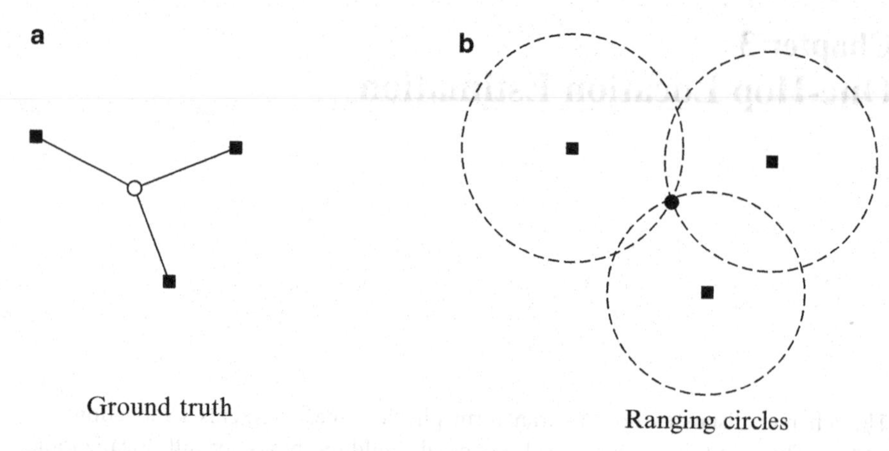

Fig. 3.1 Trilateration. (**a**) Ground truth; (**b**) ranging circles

In practice, distance measurements inevitably contain errors, resulting in that the circles may not always intersect at a single point. This problem can be solved by a numerical solution to an overdetermined linear system [31]. Suppose an unknown node locates (x_0, y_0) and it is able to obtain the distance estimates d'_i to the ith reference node locating at $(x_i, y_i), 1 \leq i \leq n$, where n is the total number of reference nodes. Let d_i be the actual Euclidean distance from the unknown node to the ith reference node, i.e.,

$$d_i = \sqrt{(x_i - x_0)^2 + (y_i - y_0)^2}.$$

Thus the difference between the measured and actual distances can be represented as $\rho_i = d'_i - d_i$. Several methods are designed to deal with the ranging noise. The least-squares method is one of them to determine the value of (x_0, y_0) by minimizing $\sum_{i=1}^{n} \rho_i^2$.

Each measurement determines an equation of the position of the unknown node, so we have

$$\begin{aligned} d_1^2 &= (x_1 - x_0)^2 + (y_1 - y_0)^2 \\ d_2^2 &= (x_2 - x_0)^2 + (y_2 - y_0)^2 \\ &\vdots \\ d_n^2 &= (x_n - x_0)^2 + (y_n - y_0)^2 \end{aligned}$$

Subtracting the first equation from all of the rest equations gives

$$
\begin{aligned}
d_2^2 - d_1^2 &= x_2^2 - x_1^2 - 2(x_2 - x_1)x_0 + y_2^2 - y_1^2 - 2(y_2 - y_1)y_0 \\
d_3^2 - d_1^2 &= x_3^2 - x_1^2 - 2(x_3 - x_1)x_0 + y_3^2 - y_1^2 - 2(y_3 - y_1)y_0 \\
&\vdots \\
d_n^2 - d_1^2 &= x_n^2 - x_1^2 - 2(x_n - x_1)x_0 + y_n^2 - y_1^2 - 2(y_n - y_1)y_0
\end{aligned}
$$

Rearranging terms, the above equations can be written in matrix form as

$$
\begin{bmatrix} x_2 - x_1 & y_2 - y_1 \\ x_3 - x_1 & y_3 - y_1 \\ \vdots & \vdots \\ x_n - x_1 & y_n - y_1 \end{bmatrix} \begin{bmatrix} x_0 \\ y_0 \end{bmatrix} = \frac{1}{2} \begin{bmatrix} x_2^2 + y_2^2 - d_2^2 - (x_1^2 + y_1^2 - d_1^2) \\ x_3^2 + y_3^2 - d_3^2 - (x_1^2 + y_1^2 - d_1^2) \\ \vdots \\ x_n^2 + y_n^2 - d_n^2 - (x_1^2 + y_1^2 - d_1^2) \end{bmatrix}
$$

Then, this equation can be rewritten as

$$
Hx = b,
$$

where

$$
H = \begin{bmatrix} x_2 - x_1 & y_2 - y_1 \\ x_3 - x_1 & y_3 - y_1 \\ \vdots & \vdots \\ x_n - x_1 & y_n - y_1 \end{bmatrix}, \quad x = \begin{bmatrix} x_0 \\ y_0 \end{bmatrix}, \quad b = \frac{1}{2} \begin{bmatrix} x_2^2 + y_2^2 - d_2^2 - (x_1^2 + y_1^2 - d_1^2) \\ x_3^2 + y_3^2 - d_3^2 - (x_1^2 + y_1^2 - d_1^2) \\ \vdots \\ x_n^2 + y_n^2 - d_n^2 - (x_1^2 + y_1^2 - d_1^2) \end{bmatrix}.
$$

The least-squares solution of this equation is given by

$$
\hat{x} = (H^{\mathrm{T}}H)^{-1}H^{\mathrm{T}}b.
$$

3.2 TDoA-Based Positioning Techniques

TDoA measurement gives the difference of the time a signal arriving at different reference nodes. A TDoA measurement Δt_{ij} and the coordinates of reference nodes i and j define one branch of a hyperbola whose foci are at the locations of reference nodes i and j. Hence, the unknown node must lie on the hyperbola. Thus, localization based on TDoA measurement is also called hyperbolic positioning. In two-dimensional space R^2, measurements from a minimum of three reference nodes are required to uniquely determine the location of an unknown node, as illustrated in Fig. 3.2.

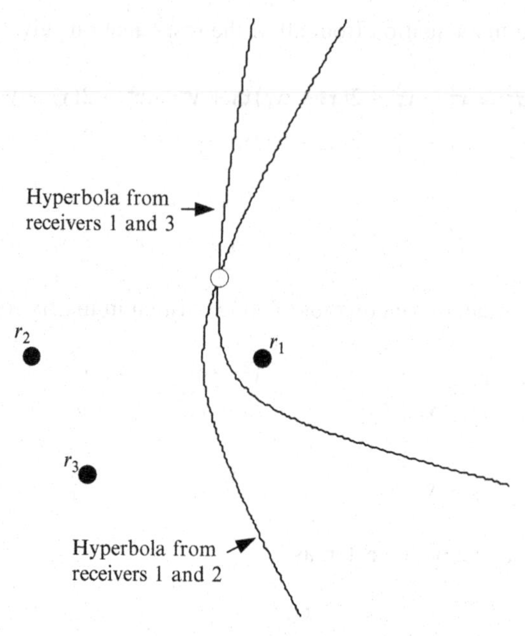

Fig. 3.2 Location computation by TDoA

Suppose we organize the TDoA measurements in the following way: the TDoA value associated with a reference node i is $\Delta t_i = t_i - t_1$, i.e., it is the difference between the arrivals of reference node 1. Let (x_0, y_0) denote the location of the unknown node, d_i denote the distance between the unknown node and reference node i, (x_i, y_i) denote the location of the reference node i. Then, we have the following basic relations:

$$\begin{aligned} d_1^2 &= (x_1 - x_0)^2 + (y_1 - y_0)^2 \\ d_2^2 &= (x_2 - x_0)^2 + (y_2 - y_0)^2 \\ &\vdots \\ d_n^2 &= (x_n - x_0)^2 + (y_n - y_0)^2 \end{aligned}$$

where n is the total number of reference nodes. Let $\Delta d_i = c\Delta t_i = d_i - d_1$, where c is the speed of the signal used by the unknown node. Then, the above equations can be rewritten as

$$\begin{aligned} d_1^2 &= (x_1 - x_0)^2 + (y_1 - y_0)^2 \\ (d_1 + \Delta d_2)^2 &= (x_2 - x_0)^2 + (y_2 - y_0)^2 \\ &\vdots \\ (d_1 + \Delta d_n)^2 &= (x_n - x_0)^2 + (y_n - y_0)^2 \end{aligned}$$

Subtracting the first equation from all of the rest equations gives

$$-(x_2 - x_1)x_0 - (y_2 - y_1)y_0 = \Delta d_2 d_1 + \frac{1}{2}(\Delta d_2^2 - x_2^2 - y_2^2 + x_1^2 + y_1^2)$$
$$-(x_3 - x_1)x_0 - (y_3 - y_1)y_0 = \Delta d_3 d_1 + \frac{1}{2}(\Delta d_3^2 - x_3^2 - y_3^2 + x_1^2 + y_1^2)$$
$$\vdots$$
$$-(x_n - x_1)x_0 - (y_n - y_1)y_0 = \Delta d_n d_1 + \frac{1}{2}(\Delta d_n^2 - x_n^2 - y_n^2 + x_1^2 + y_1^2)$$

Rewriting these equations in matrix form gives

$$Hx = d_1 a + b,$$

where

$$H = \begin{bmatrix} x_2 - x_1 y_2 - y_1 \\ x_3 - x_1 y_3 - y_1 \\ \vdots \\ x_n - x_1 y_n - y_1 \end{bmatrix}, x = \begin{bmatrix} x_0 \\ y_0 \end{bmatrix}, a = \begin{bmatrix} -\Delta d_2 \\ -\Delta d_3 \\ \vdots \\ -\Delta d_n \end{bmatrix}, b = -\frac{1}{2}\begin{bmatrix} \Delta d_2^2 - x_2^2 - y_2^2 + x_1^2 + y_1^2 \\ \Delta d_3^2 - x_3^2 - y_3^2 + x_1^2 + y_1^2 \\ \vdots \\ \Delta d_n^2 - x_n^2 - y_n^2 + x_1^2 + y_1^2 \end{bmatrix}.$$

The least-squares estimation of this equation is given by

$$\hat{x} = (H^{\mathrm{T}}H)^{-1}H^{\mathrm{T}}(d_1 a + b).$$

In this result, parameter d_1 is unknown. Note that we have ${d_1}^2 = (x_1 - x_0)^2 + (y_1 - y_0)^2$. Substituting the above intermediate result into this equation leads to a quadratic equation of d_1. Solving for d_1 and substituting the positive root back into the least-squares estimation yields the final solution for x, i.e., the location estimate of the unknown node.

Other than the basic least-squares solution, researchers have developed several techniques to solve the nonlinear equations of TDoA localization [50–52]. Being accurate and robust, the Taylor-series method [50] is commonly used to deal with nonlinearity. It is an iterative method under the prerequisite that the initial guess is close to the true solution to avoid local minima. However, the selection of such a starting point is not simple in practice. Using least-squares estimation two times, Chan [52] propose a closed form, noniterative solution, which performs well when the TDoA measurement errors are small. However, as the errors increase, the performance degrades quickly.

3.3 AoA-Based Positioning Techniques

AoA measurement gives the bearing information of two nodes, as shown in Fig. 3.3. Let (x_0, y_0) be the location of the unknown node to be estimated from AoA measurement $\alpha_i, 1 \leq i \leq n$, where n is the total number of reference nodes.

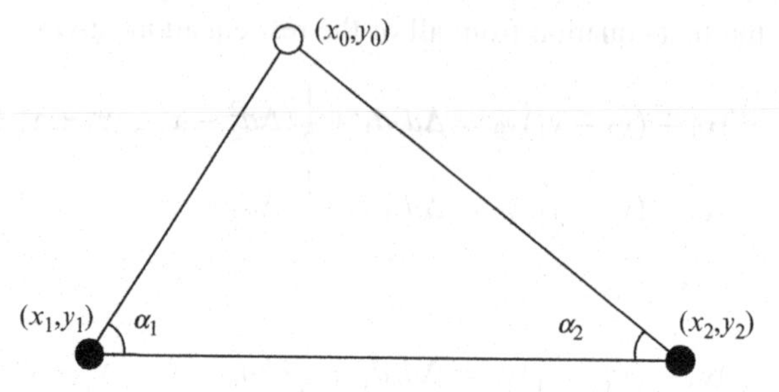

Fig. 3.3 Location computation by AoA measurement

Let (x_i, y_i) be the known location of the reference node i, $\theta_i(p)$ denote the bearing of a node located at $\vec{x} = (x, y)$. We have

$$\tan\theta_i(\vec{x}) = \frac{y - y_i}{x - x_i}, 1 \leq i \leq n.$$

Suppose the measured bearings of reference node i are corrupted by additive noises $\varepsilon_i, 1 \leq i \leq n$, which are assumed to be zero-mean Gaussian noises with covariance matrices ${\sigma_i}^2$, i.e.,

$$\alpha_i = \theta_I(x_0) + \epsilon_i, 1 \leq i \leq n.$$

When the reference nodes are identical and much closer to each other than to the unknown node, the variances of bearing measurement errors are equal, i.e., $\sigma_i^2 = \sigma^2, 1 \leq i \leq n$. The maximum likelihood estimator of the location of the unknown node is given by

$$\hat{x} = \arg\min \frac{1}{2} \sum_{i=1}^{n} \frac{(\theta_i(\hat{x}) - \alpha_i)^2}{\sigma_i^2}$$

This nonlinear minimization problem can be solved by a Newton–Gauss iteration [53].

Another approach bases on the assumption that the measurement error is small enough such that $\varepsilon_i \approx \sin(\varepsilon_i)$. In that case, the above cost function becomes

$$\frac{1}{2} \sum_{i=1}^{n} \frac{\sin^2(\theta_i(\hat{x}) - \alpha_i)}{\sigma_i^2}.$$

According to $d_i = \sqrt{(x_0 - x_i)^2 + (y_0 - y_i)^2}$ and

$$\begin{aligned}\sin(\theta_i(\hat{x}) - \alpha_i) &= \sin\theta_i(\hat{x})\cos\alpha_i - \cos\theta_i(\hat{x})\sin\alpha_i \\ &= \frac{(y_0 - y_i)\cos\alpha_i - (x_0 - x_i)\sin\alpha_i}{d_i},\end{aligned}$$

the cost function becomes

$$\begin{aligned}&\frac{1}{2}\sum_{i=1}^{n}\frac{[(y_0 - y_i)\cos\alpha_i - (x_0 - x_i)\sin\alpha_i]^2}{\sigma_i^2 d_i^2} \\ &= \frac{1}{2}(A\underline{x} - b)^T R^{-1} S^{-1}(A\underline{x} - b),\end{aligned}$$

where

$$A = \begin{bmatrix}\sin\alpha_1 & -\cos\alpha_1 \\ \vdots & \vdots \\ \sin\alpha_n & -\cos\alpha_n\end{bmatrix},$$

$$b = \begin{bmatrix}x_1\sin\alpha_1 - y_1\cos\alpha_1 \\ \vdots \\ x_n\sin\alpha_n - y_n\cos\alpha_n\end{bmatrix},$$

$$R = \mathrm{diag}\{d_1^2, \ldots, d_n^2\},$$

$$S = \mathrm{diag}\{\sigma_1^2, \ldots, \sigma_n^2\}.$$

This method implicitly assumes that a rough estimate of R can be obtained. Since the cost function weakly depends on R, the roughness will not significantly affect the solution. Under these assumptions, the minimum cost solution with respect to $\vec{x}$ is given by

$$\hat{x} = (A^{\mathrm{T}}R^{-1}S^{-1}A)^{-1}A^{\mathrm{T}}R^{-1}S^{-1}b.$$

3.4 RSS-Profiling-Based Positioning Techniques

RSS-profiling-based positioning techniques directly utilize RSS data for location estimation. In indoor environments, mapping RSS to distance measurement may introduce huge errors, because RSS is strongly affected by the shadowing and multipath effect. However, the RSS distribution of a set of anchor nodes is relatively stable over the spatial space, so the RSS vector measured by an unknown node, defined as RSS finger print, reveals the physical location of the node. By contrasting the RSS finger print with the profiled data, the location of the unknown node is estimated. Based on the schemes of profiling, existing approaches fall in to two categories: off-line profiling and online profiling.

3.4.1 Off-line Profiling Scheme

A typical off-line profiling scheme is RADAR [29], which positions an unknown node by building an RSS-location map. RADAR contains two steps: off-line map sensing and online node positioning. In the first step, RADAR collects the spatial distribution of the RSS of the anchors to build a RSS-location map. Specifically, system operators in advance conduct a site survey by recording the RSS values at each location in an interesting area. The RSS at a given location varies quite significantly (by up to 5 dBm) depending on the user's orientation, i.e., the direction he/she is facing. Hence, RADAR takes into account the direction and records the following tuple at each sample point (t,x,y,d), where t denotes the timestamp of the measurement, (x,y) and d show the position and direction of the measurement, respectively.

After building the RSS-location-direction map, RADAR can provide online positioning service, which is the second step. Each unknown node first measures the RSS between the anchor nodes within its communication range, and thus creates its own RSS finger print. Then, it transmits the RSS finger print to the central station. Using this RSS finger print, the central station matches the presented signal strength vector to the RSS-location-direction map, using the nearest-neighbor-based method. That is, the location of a sample point, whose RSS vector is the closest match to that of the unknown node, is chosen to be the estimated location of the nonanchor node.

Besides the merit of simplicity, RADAR can also properly handle the mobility of the unknown node. However, the accuracy of such scheme suffers the environmental dynamics.

3.4.2 Online Profiling Scheme

The off-line map for RSS-profiling suffers the environmental dynamics, which is a main characteristic of the wireless communication. One way to address this issue is to use the online map for positioning the unknown nodes, called LANDMARC [54]. LANDMARC design is based on the radio frequency identification (RFID) technology, which is a means of storing and retrieving data through electromagnetic transmission to an RF-compatible integrated circuit. An RFID system has several basic components including a number of RFID readers and RFID tags. The RFID reader can read data emitted from RFID tags. RFID readers and tags use a defined radio frequency and protocol to transmit and receive data. RFID tags are categorized as either passive or active. Passive RFID tags operate without a battery. Active tags contain both a radio transceiver and a button cell battery to power the transceiver. Since there is an onboard radio on the tag, active tags have larger range than passive tags.

Positioning based on the online map does not need to collect the RSS distribution prior. LANDMARC employs the idea of exploiting extra fixed location reference tags to help location calibration. These reference tags serve as reference points in the system (like landmarks in our daily life). The advantage of this design is to achieve high localization accuracy from the cost of tags instead of the readers, because the RFID readers are much more expensive than the RFID tags.

The reference tags forms an online map for location computation. As shown in Fig. 3.4, the predeployed reference tags cover the target area well and uniformly provide sample data to locate the tracking tags. Note that the RF readers can read all tags in the target area, including the reference tags and the tracking tags.

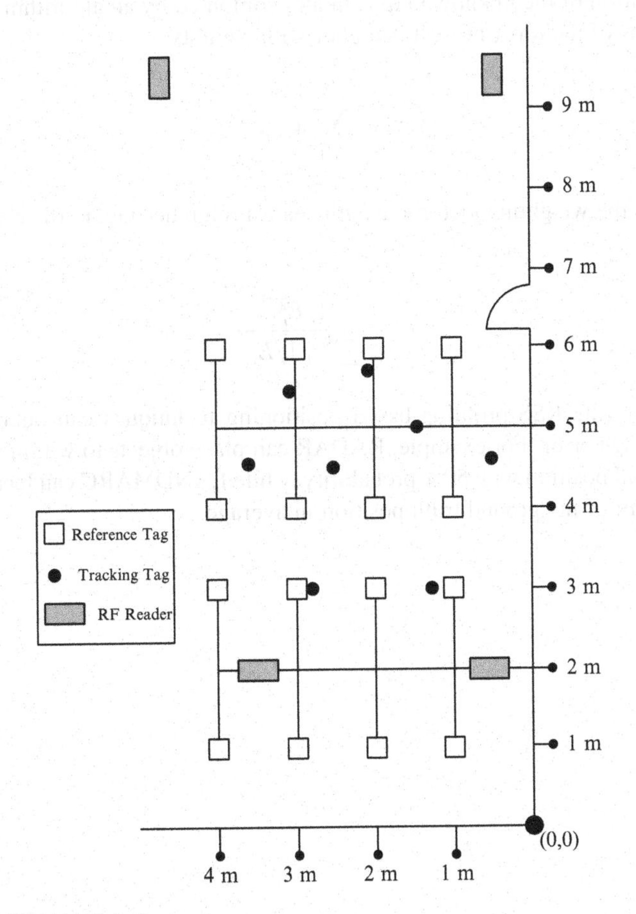

Fig. 3.4 LANDMARC deployment

The location computation of a tracking tag is as follows. Suppose there are n RF readers along with m tags as reference tags and u tracking tags as objects being tracked. The readers are all configured with continuous mode (continuously reporting the tags that are within the specified range) and a detection range of 1–8 (meaning the reader will scan from range 1 to 8 and keep repeating the cycle with a rate of 30 s per range). Define the signal strength vector of a tracking/moving tag as $S = (S_1, S_2, \ldots, S_n)$, where S_i denotes the signal strength of the tracking tag perceived on reader i, $i \in (1, n)$. For the reference tags, let $\theta = (\theta_1, \theta_2, \ldots, \theta_n)$ denote the corresponding signal strength vector, where θ_i denotes the signal strength. LANDMARC adopts the Euclidean distance in signal strengths. For each individualtracking tag p, $p \in (1, u)$, define $E_j = \sqrt{\sum_{i=1}^{n} (\theta_i - S_i)^2}$, $j \in (1, m)$, as the Euclidean distance in signal strength between a tracking tag and a reference tag r_j. Let E denote the location relationship between the reference tags and the tracking tag, i.e., the nearer reference tag to the tracking tag is supposed to have a smaller E value. When there are m reference tags, a tracking tag has its E vector as $E = (E_1, E_2, \ldots, E_m)$.

The location of the unknown tag is finally computed by an algorithm averaging the positions of the top k nearest neighbor with weights

$$(x, y) = \sum_{i=1}^{k} w_i (x_i, y_i),$$

where w_i is the weighting factor to the ith nearest reference tag. Further, the weight is given by

$$w_i = \frac{1/E_i^2}{\sum_{i=1}^{k} 1/E_i^2}.$$

Generally, the RSS-profiling-based positioning techniques can obtain several meters average error. For example, RADAR can place objects to within about 3 m of their actual position with 50% probability, while LANDMARC can localize a tag to within 1 m of the ground truth position in average.

Chapter 4
Range-Based Network localization

4.1 Computation Organization

This section defines the taxonomy of range-based approaches based on their computational organization. Centralized algorithms are designed to run on a central machine with plenty of resources. Network nodes collect physical measurements and deliver back to a base station for analysis. Centralized algorithms resolve the computational limitations of individual nodes. This benefit, however, comes from accepting the communication cost of transmitting data back to the base station. Unfortunately, communication generally consumes more energy than computation in most hardware platforms.

In contrast, distributed algorithms are designed to run in-network, using massive parallelism and internode communication to compensate for the lack of centralized computing power, while at the same time reducing the expensive node-to-sink communication. Often distributed algorithms use a subset of measurement data to locate nodes one by one, yielding an approximation of a corresponding centralized algorithm where all the data are considered and used to compute the positions of all nodes simultaneously. There are two important categories of distributed approaches. The first group, beacon-based distributed algorithms, typically starts a localization process with beacons and the nodes in vicinity of beacons. In general, nodes measure distances to a few beacons and then determine their locations. In some algorithms, the newly localized nodes become beacons to help locating other nodes in the following process. In such an iterative fashion, location information diffuses from beacons to the border of a network, which can be viewed as a top-down manner. The second group of approaches performs in a bottom-up manner, in which localization is originated in a local group of nodes in relative coordinates. After gradually merging such local maps, it finally achieves entire network localization in global coordinates.

Y. Liu and Z. Yang, *Location, Localization, and Localizability: Location-awareness Technology for Wireless Networks*, DOI 10.1007/978-1-4419-7371-9_4,

4.2 Centralized Localization Approaches

4.2.1 Multidimensional Scaling (MDS)

Multidimensional scaling (MDS) [55] is originally developed for mathematical psychology. The intuition behind MDS is simple. Suppose there are n points, suspended in a volume. We do not know the positions of the points, but we know the distance between each pair of points. MDS is an $O(n^3)$ algorithm that uses the law of cosines and linear algebra to reconstruct the relative positions of the points based on the pairwise distances. The algorithm has three stages:

1. Generate an $n \times n$ matrix M, whose (i, j) entry contains the estimated distance between nodes i and j. (Simply run Floyd's all-pairs shortest-path algorithm.)
2. Apply classical metric MDS on M to determine a map that gives the locations of all nodes in relative coordinates.
3. Transform the solution into global coordinates based on a number of anchor nodes.

The goal of metric MDS is to find a configuration of points in a multidimensional space such that the interpoint distances are related to the provided proximities by some transformation (e.g., a linear transformation). The computation of metric MDS is as follows.

Let p_{ij} refer to the proximity measure between objects i and j. The Euclidean distance between two points $X_i = (x_{i1}, x_{i2}, \ldots, x_{im})$ and $X_j = (x_{j1}, x_{j2}, \ldots, x_{jm})$ in an m-dimensional space is given by

$$d_{ij} = \sqrt{\sum_{k=1}^{m} (x_{ik} - x_{jk})^2}.$$

When a geometrical model (the coordinates of points) fits the proximity data M perfectly, the corresponding Euclidean distances are related to the proximities by a transformation $d_{ij} = f(p_{ij})$. In classical metric MDS, a linear transformation model is assumed, i.e., $d_{ij} = a + bp_{ij}$.

The distances D are determined so that they are as close to the proximities P as possible, under a least-squares metric. In this case, define $I(P) = D + E$, where $I(P)$ is a linear transformation of the proximities and E is a matrix of errors (residuals). Since D is a function of the coordinates X, the goal of classical metric MDS is to calculate the X such that the sum of squares of E is minimized, subject to suitable normalization of X. In classical metric MDS, P is shifted to the center and coordinates X can be computed from the double centered P through singular value decomposition (SVD). For an $n \times n$ P matrix for n points and m dimensions of each point, we have

$$-\frac{1}{2}\left(p_{ij}^2 - \frac{1}{n}\sum_{j=1}^{n} p_{ij}^2 - \frac{1}{n}\sum_{i=1}^{n} p_{ij}^2 + \frac{1}{n^2}\sum_{i=1}^{n}\sum_{j=1}^{n} p_{ij}^2\right) = \sum_{k=1}^{m} x_{ik}x_{jk}.$$

The double-centered matrix on the left-hand side (call it B) is symmetric and positive semidefinite. Performing singular value decomposition on B gives $B = VAV$. The coordinate matrix becomes $X = VA^{1/2}$.

Retaining the first r largest eigenvalues and eigenvectors ($r < m$) leads to a solution for lower dimensions. This implies that the summation over k in the above equation runs from 1 to r instead of m. It is the best low-rank approximation in the least-squares sense. For example, for a 2D network, we take the first two largest eigenvalues and eigenvectors to construct the best 2D approximation; while the first three largest ones for 3D case.

MDS performs well on RSS data, getting performance on the order of half the radio range when the neighborhood size n_{local} is higher than 12 [56]. The main problem with MDS, however, is its poor asymptotic performance, which is $O(n^3)$ on account of stages 1 and 2.

Besides the computation cost, the classical metric MDS has the other two main drawbacks. First, the computation is inherently centralized, which constrains the scalability of MDS. Second, for irregularly shaped networks, the shortest path distance between two nodes does not correspond well to their Euclidean distance. Consequently, the distance estimation error will introduce huge errors in the localization result. To address the problem, researchers propose a distributed MDS-based algorithm, called MDS-MAP(P) [57]. The main idea of MDS-MAP (P) is to build a local map at each node of the immediate vicinity and then merge these maps together to form a global map. Specifically, MDS-MAP(P) includes the following five steps:

1. Set the range for local maps, R_{lm}. For each node, neighbors within R_{lm} hops are involved in building its local map. The value of R_{lm} affects the amount of computation, as well as the quality. Generally, setting $R_{lm} = 2$ can obtain satisfactory results. The overall complexity of computing each local map is $O(k^3)$, where k is the average number of neighbors. Thus the complexity of computing n local maps is $O(k^3 n)$, where n is the number of nodes.
2. Compute local maps for individual nodes. For each node, do the following:
   (a) Compute shortest paths between all pairs of nodes in its local mapping range R_{lm}. The shortest path distances are used to construct the distance matrix for MDS.
   (b) Apply MDS to the distance matrix and retain the first two (or three) largest eigenvalues and eigenvectors to construct a 2D (or 3D) local map.
   (c) Refine the local map. Using the node coordinates in the MDS solution as the initial point, least-squares minimization is performed to make the distances between nearby nodes match the measured ones.
3. *Merge local maps*. Local maps can be merged sequentially or in parallel. First, randomly pick a node and make its local map the core map. Then, grow the core

map by merging maps of neighboring nodes to the core map. Each time a neighbor's map with the maximal number of common nodes with the core map is selected. Eventually the core map covers the whole network. If the merges are chosen carefully, the complexity of this step is $O(k^3 n)$, where k is the average number of neighbors and n is the number of nodes.
4. *Refine the global map (optional).* Using the node coordinates in the global map as the initial solution, least-squares minimization is applied to make the distances between neighboring nodes match the measured ones. This step leads to $O(n^3)$ cost.
5. Given sufficient anchor nodes (three or more for 2D networks, four or more for 3D networks), transform the global map to an absolute map based on the absolute positions of anchors. For r anchors, the complexity of this step is $O(r^3 + n)$.

To summarize, MDS-MAP(P) computes small relative maps using local information, instead of a global map using pairwise distances between any two nodes. Comparing with the centralized version, MDS-MAP(P) reduces the computational complexity and can handle the nonconvexity of network deployment, thus the localization accuracy is improved.

4.2.2 Semidefinite Programming (SDP)

Semidefinite programming (SDP) is pioneered by Doherty [39]. In this algorithm, geometric constraints between nodes are represented as linear matrix inequalities (LMIs). Once all the constraints in the network are expressed in this form, the LMIs can be combined to form a single semidefinite program, which is solved to produce a bounding region for each node. The advantage of SDP is its elegance on concise problem formulation, clear model representation, and elegant mathematic solution.

The mathematical expression of SDP is as follows. Suppose there are m anchors $a_k \in R^2, k = 1, \ldots, m$, and n unknown nodes $x_j \in R^2, j = 1, \ldots, n$. For a pair of two points in N_e, we have a Euclidean distance measure d_{kj} between a_k and x_j or d_{ij} between x_i and x_j; and for a pair of two points in N_l, we have a distance lower bound $\underline{r}_{kj}$ between a_k and x_j or $\underline{r}_{ij}$ between x_i and x_j; and for a pair of two points in N_u, we have a distance upper bound $\bar{r}_{kj}$ between a_k and x_j or $\bar{r}_{ij}$ between x_i and x_j. Then, the localization problem is to find x_js such that

$$\begin{aligned}
&\| x_i - x_j \|^2 = d_{ij}^2, \| a_k - x_j \|^2 = d_{kj}^2, \forall (i,j), (k,j) \in N_e \\
&\| x_i - x_j \|^2 \geq \underline{r}_{ij}^2, \| a_k - x_j \|^2 \geq \underline{r}_{kj}^2, \forall (i,j), (k,j) \in N_l \\
&\| x_i - x_j \|^2 \leq \bar{r}_{ij}^2, \| a_k - x_j \|^2 \leq \bar{r}_{kj}^2, \forall (i,j), (k,j) \in N_u
\end{aligned}$$

Since these measures and bounds are typically noisy, the coordinates x_js are chosen to minimize the sum of errors:

$$
\begin{aligned}
\min & \sum_{i,j\in N_e,i<j} \left| \| x_i - x_j \|^2 - d_{ij}^2 \right| \\
& + \sum_{k,j\in N_e} \left| \| a_k - x_j \|^2 - d_{kj}^2 \right| \\
& + \sum_{i,j\in N_l,i<j} (\| x_i - x_j \|^2 - \underline{r}_{ij}^2)_- \\
& + \sum_{k,j\in N_l} (\| a_k - x_j \|^2 - \underline{r}_{kj}^2)_- \\
& + \sum_{i,j\in N_u,i<j} (\| x_i - x_j \|^2 - \bar{r}_{ij}^2)_+ \\
& + \sum_{k,j\in N_u} (\| a_k - x_j \|^2 - \bar{r}_{kj}^2)_+,
\end{aligned}
$$

where $(u)_-$ and $(u)_+$ are defined as $(u)_- = \max\{0, -u\}$ and $(u)_+ = \max\{0, u\}$, respectively.

Let $X = [x_1\ x_2 \cdots x_n]$ be the $2 \times n$ matrix that needs to be determined. Then

$$\| x_i - x_j \|^2 = e_{ij}^{\mathrm{T}} X^{\mathrm{T}} X e_{ij},$$

$$\| a_i - x_j \|^2 = (a_i; e_j)^{\mathrm{T}} [I\ X]^{\mathrm{T}} [I\ X] (a_i; e_j,)$$

where e_{ij} is a $n \times 1$ vector with 1 at the ith position, -1 at the jth position, and zeros elsewhere; and e_j is the vector of all zero except -1 at the jth position. Let $Y = X^{\mathrm{T}} X$. By introducing slack variables αs and βs, the softer error minimization problem can be rewritten as

$$
\begin{aligned}
\min \quad & \sum_{i,j\in N_e,i<j} (\alpha_{ij}^+ + \alpha_{ij}^-) + \sum_{k,j\in N_e} (\alpha_{kj}^+ + \alpha_{kj}^-) \\
& + \sum_{i,j\in N_l,i<j} \beta_{ij}^- + \sum_{k,j\in N_l} \beta_{kj}^- \\
& + \sum_{i,j\in N_u,i<j} \beta_{ij}^+ + \sum_{k,j\in N_u} \beta_{ij}^+ \\
\text{s.t.} \quad & e_{ij}^{\mathrm{T}} Y e_{ij} - d_{ij}^2 = \alpha_{ij}^+ - \alpha_{ij}^-, \forall i,j \in N_e, i<j, \\
& (a_k; e_j)^{\mathrm{T}} \begin{pmatrix} IX \\ X^T Y \end{pmatrix} (a_k; e_j) - d_{kj}^2 = \alpha_{kj}^+ - \alpha_{kj}^-, \forall k,j \in N_e, \\
& e_{ij}^{\mathrm{T}} Y e_{ij} - \underline{r}_{ij}^2 \geq -\beta_{ij}^-, \forall i,j \in N_l, i<j, \\
& (a_k; e_j)^{\mathrm{T}} \begin{pmatrix} IX \\ X^{\mathrm{T}} Y \end{pmatrix} (a_k; e_j) - \underline{r}_{kj}^2 \geq -\beta_{kj}^-, \forall k,j \in N_l, \\
& e_{ij}^{\mathrm{T}} Y e_{ij} - \bar{r}_{ij}^2 \leq \beta_{ij}^+, \forall i,j \in N_u, i<j, \\
& (a_k; e_j)^{\mathrm{T}} \begin{pmatrix} IX \\ X^{\mathrm{T}} Y \end{pmatrix} (a_k; e_j) - \bar{r}_{kj}^2 \leq \beta_{kj}^+, \forall k,j \in N_u, \\
& \alpha_{ij}^+, \alpha_{ij}^-, \alpha_{kj}^+, \alpha_{kj}^-, \beta_{ij}^-, \beta_{kj}^-, \beta_{ij}^+, \beta_{kj}^+ \geq 0, \\
& Y = X^{\mathrm{T}} X.
\end{aligned}
$$

Unfortunately, the above problem is not a convex optimization problem. Biswas et al. in IPSN'04 propose to convert this problem to a semidefinite program, by relaxing $Y = X^{\mathrm{T}}X$ to $Y \succeq X^{\mathrm{T}}X$. The expression $Y \succeq X^{\mathrm{T}}X$ is equivalent to

$$Z := \begin{pmatrix} I & X \\ X^{\mathrm{T}} & Y \end{pmatrix} \succeq 0.$$

Then, the problem can be written as a standard SDP problem:

$$\begin{aligned}
\min \quad & \sum\nolimits_{i,j \in N_e, i<j} (\alpha_{ij}^{+} + \alpha_{ij}^{-}) + \sum\nolimits_{k,j \in N_e} (\alpha_{kj}^{+} + \alpha_{kj}^{-}) \\
& + \sum\nolimits_{i,j \in N_l, i<j} \beta_{ij}^{-} + \sum\nolimits_{k,j \in N_l} \beta_{kj}^{-} \\
& + \sum\nolimits_{i,j \in N_u, i<j} \beta_{ij}^{+} + \sum\nolimits_{k,j \in N_u} \beta_{ij}^{+} \\
\text{s.t.} \quad & (1;0;0)^{\mathrm{T}} Z (1;0;0) = 1 \\
& (0;1;0)^{\mathrm{T}} Z (0;1;0) = 1 \\
& (1;1;0)^{\mathrm{T}} Z (1;1;0) = 2 \\
& (0;e_{ij})^{\mathrm{T}} Z (0;e_{ij}) - \alpha_{ij}^{+} + \alpha_{ij}^{-} = d_{ij}^{2}, \forall i,j \in N_e, i<j, \\
& (a_k;e_j)^{\mathrm{T}} Z (a_k;e_j) - \alpha_{kj}^{+} + \alpha_{kj}^{-} = d_{kj}^{2}, \forall k,j \in N_e, \\
& (0;e_{ij})^{\mathrm{T}} Z (0;e_{ij}) + \beta_{ij}^{-} \geq \underline{r}_{ij}^{2}, \forall i,j \in N_l, i<j, \\
& (a_k;e_j)^{\mathrm{T}} Z (a_k;e_j) + \beta_{kj}^{-} \geq \underline{r}_{kj}^{2}, \forall k,j \in N_l, \\
& (0;e_{ij})^{\mathrm{T}} Z (0;e_{ij}) - \beta_{ij}^{+} \leq \bar{r}_{ij}^{2}, \forall i,j \in N_u, i<j, \\
& (a_k;e_j)^{\mathrm{T}} Z (a_k;e_j) - \beta_{kj}^{+} \leq \bar{r}_{kj}^{2}, \forall k,j \in N_u, \\
& \alpha_{ij}^{+}, \alpha_{ij}^{-}, \alpha_{kj}^{+}, \alpha_{kj}^{-}, \beta_{ij}^{-}, \beta_{kj}^{-}, \beta_{ij}^{+}, \beta_{kj}^{+} \geq 0, \\
& Z \succeq 0.
\end{aligned}$$

Solving the linear or semidefinite program centrally, the time complexity is $O(k^2)$ for angle of arrival data, and $O(k^3)$ when radial (e.g., hop count) data are included, where k is the number of convex constraints needed to describe a network. Thus, the computation complexity of SDP is likely to preclude itself in practice.

4.3 Distributed Localization Approaches

4.3.1 Beacon-based Localization

4.3.1.1 Iterative Trilateration

Beacon-based localization approaches utilize the node-to-beacon distances. The distance between an unknown node and a beacon can be estimated using a basic

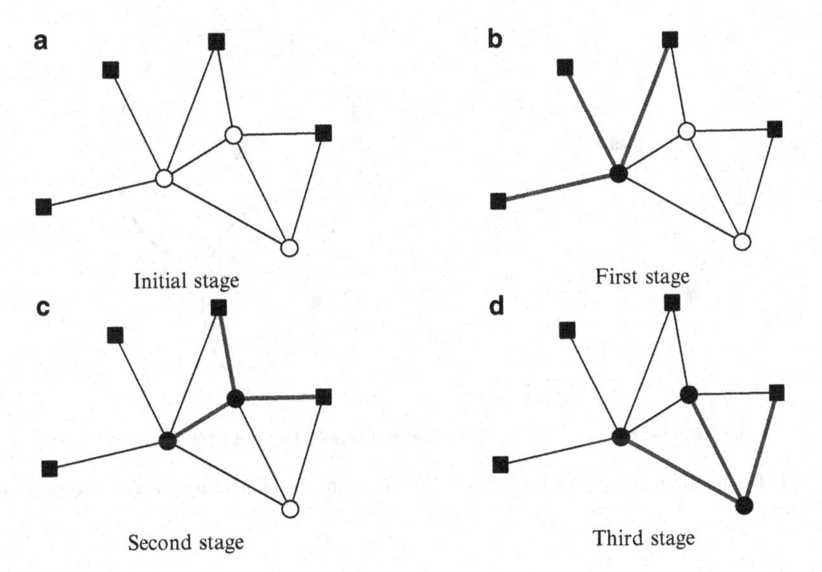

Fig. 4.1 Iterative localization. Bring a to d here (**a**) Initial stage; (**b**) first stage; (**c**) second stage; (**d**) third stage

distance-vector technique [58, 59]. Such a mechanism can be viewed as a top-down manner due to the progressive propagation of location information from beacons to entire networks.

One variant of this approach is the indirect usage of beacon nodes. Initially an unknown node, if possible, is located based on its neighbors by multilateration or other positioning techniques. After being aware of its location, it becomes a reference node to localize other unknown nodes in the subsequent process. This step continues iteratively, gradually changing unknown nodes to known ones. The process of iterative trilateration is illustrated in Fig. 4.1, in which squares are beacons, soft circles are unknown nodes, and solid circles are known nodes.

The advantage of this approach is that it only involves local information (information within neighborhood) when locating a node, leading to high efficiency in terms of communication. However, the use of localized unknown nodes as reference inherently introduces substantial cumulative errors, especially for the nodes far away from beacons. Some works [60, 61] characterize the error propagation in multihop localization approaches and make efforts to control error accumulation. The error control techniques will be discussed in detail in Chap. 6.

4.3.1.2 Finite Localization by Bilateration

Experimental studies show that trilateration-based algorithms require an average node degree beyond 10 for correctly localizing most of the nodes in a network [62]. When the average degree is below 8, the iterative trilateration will fail in most

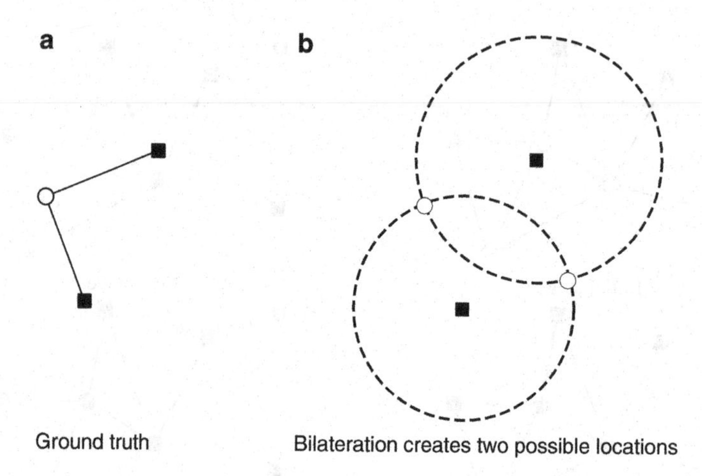

Fig. 4.2 Bilateration. Bring a and b here (**a**) Ground truth; (**b**) bilateration creates two possible locations

cases. To be more applicable for sparse networks, sweeps [63] partially relaxes the requirement of node dependence in iterative approaches. In contrast to the traditional "unique position computation," sweeps presents finite localization which locates a target node to a set of possible positions called candidate positions. Finite localization guarantees that the ground truth position of a node is one of its candidate positions. Further, sweeps adopts a new positioning scheme called bilateration, which computes the candidate positions of a node by utilizing the distance measurements of only two reference nodes. As shown in Fig. 4.2, bilateration produces two candidate positions (soft dots) for an unknown node and one of them (the left one) is the ground-truth position. Similar to multilateration, the finitely localized node, called swept node, can act as a reference node to localize other unknown nodes. The only difference is that all candidate positions of the swept node are enumerated for the location computation of the target node. Moreover, after each bilateration, sweeps checks the consistency among the candidate position sets and deletes the incompatible items. Under this mechanism, sweeps can locate a large proposition of theoretically localizable nodes in a network. However, the worst case computation grows exponentially in terms of the number of nodes.

The details of sweeps design are demonstrated through a typical network topology, as shown in Fig. 4.3 a, in which solid dots denote reference nodes and soft dotes denote unknown nodes. Clearly, traditional trilateration-based algorithms cannot localize any of the unknown nodes. In contrast, sweeps can locate v_4 that has the distance measurements to two swept nodes v_1 and v_3. This bilateration generates two candidate positions, as shown in Fig. 4.3 b1, b2. As node v_4 is finitely localized, it becomes a swept node. Then, node v_5 knows the distances to two swept nodes v_1 and v_4, so it can be finitely localized. Note that, all candidate positions of node v_4 are enumerated to compute the candidate positions of node v_5. Based on the two

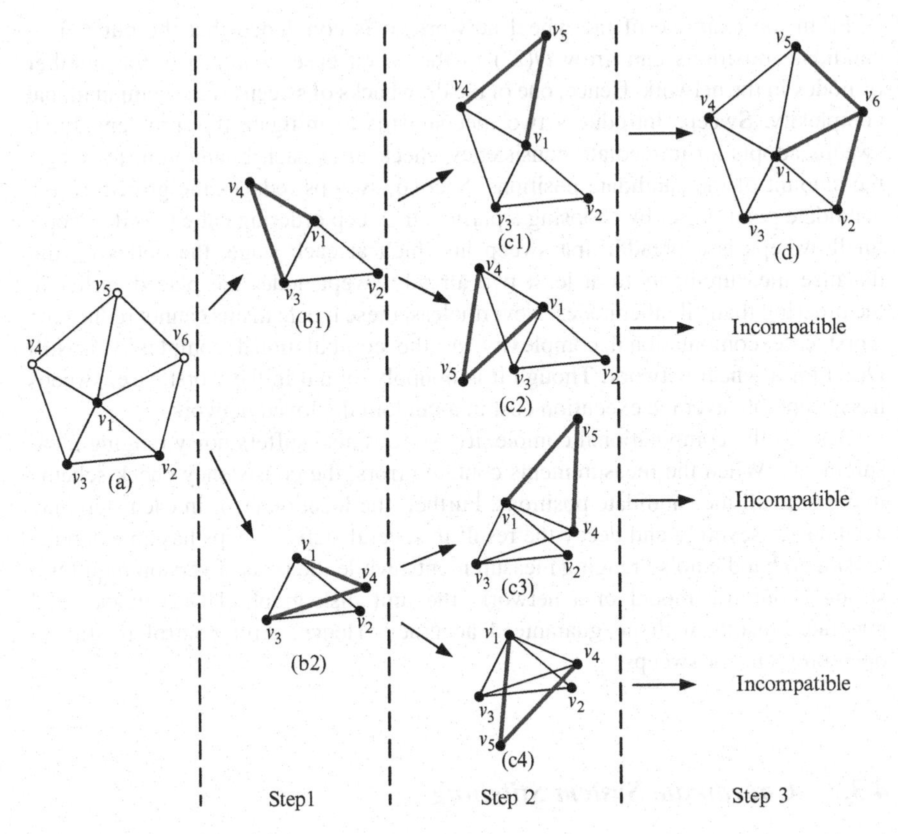

Fig. 4.3 Sweeps execution

candidate positions of node v_4, node v_5 has four candidate positions. The dependence relationship of the candidate positions is shown by arrows in Fig. 4.3. All the candidate positions of node v_5 are consistent with the distant measurements, as shown in Fig. 4.3 c1–c4. Finally, node v_6 has distance measurements to three swept nodes v_1, v_2, and v_5. Due to the consistency check of these distance measurements, only one of the candidate positions of node v_5 is compatible. Hence, all the incompatible candidate positions of node v_5 are pruned. Further, the related candidate positions (by the dependence tree) of node v_4 are also pruned. Eventually, all the nodes in this network are properly localized.

Sweeps can localize a kind of network called localizable bilateration network, formally defined as follows. A network has a bilateration ordering with anchors v_1, v_2, and v_3 if its nodes can be ordered as $v_1, v_2, \ldots, v_n$ so that v_1 and v_2 are adjacent, and each v_i, $i > 2$, is adjacent to at least two vertices v_j where $j < i$. A network is called a bilateration network if it has a bilateration ordering. Clearly, the wheel network illustrated in Fig. 4.3 a is a special case of the bilateration network.

From the example of the wheel network, it is concluded that the number of candidate positions can grow $O(2^n)$ in the worst case, where n is the number of nodes in the network. Hence, one of the drawbacks of sweeps is the computational complexity. Sweeps introduces two mechanisms to mitigate this problem. First, sweeps adopts an immediate consistency check after each bilateration to reduce the amount of the candidate positions. Second, sweeps reduces the growth of the candidate position set by choosing a particular sweep ordering called shell sweeps. Shell sweeps is a breadth-first sweep in which at each stage, the nodes having distance measurements to at least two already swept nodes are placed earlier in the ordering than all other nodes. Nevertheless, these mechanisms cannot reduce the worst case computational complexity, as the computational complexity is still $O(2^n)$ for a wheel network. Though it is nonpolynomial in the worst case, sweeps has acceptable average execution cost in a random deployed network.

Besides the computational complexity, sweeps also suffers noisy ranging measurements. When the measurements contain errors, the consistency check scheme may prune all the candidate positions. Further, the location error in each step may accumulate severely and decay the result in several steps. sweeps has an extended version to handle noisy ranging measurements, while the revised version requires a strong geometric model of a network, the unit disk graph (UDG) model, and generates results with no guaranteed accuracy. Hence, error control is still an open problem for sweeps.

4.3.2 Coordinate System Stitching

4.3.2.1 Local Map Stitching

Coordinate system stitching is an alternative for localization and has attracted a lot of research efforts recently [58, 62, 64]. It works in a bottom-up manner, in which localization is originated in a group of local nodes in relative coordinates. By gradually merging local maps, it finally achieves entire network localization in global coordinates, illustrated in Fig. 4.4.

Coordinate system stitching typically works as follows:

1. Split the network into small overlapping subregions. Very often each subregion is simply a single node and its one-hop neighbors.
2. For each subregion, compute a "local map," which is essentially an embedding of the nodes in the subregion into a relative coordinate system.
3. Finally, merge subregions using a coordinate system registration procedure. Coordinate system registration finds a rigid transformation that maps points in one coordinate system to a different coordinate system. Thus, step 3 places all the subregions into a single global coordinate system. Many algorithms do this step suboptimally, since there is a closed-form, fast and least-squares optimal method of registering coordinate system.

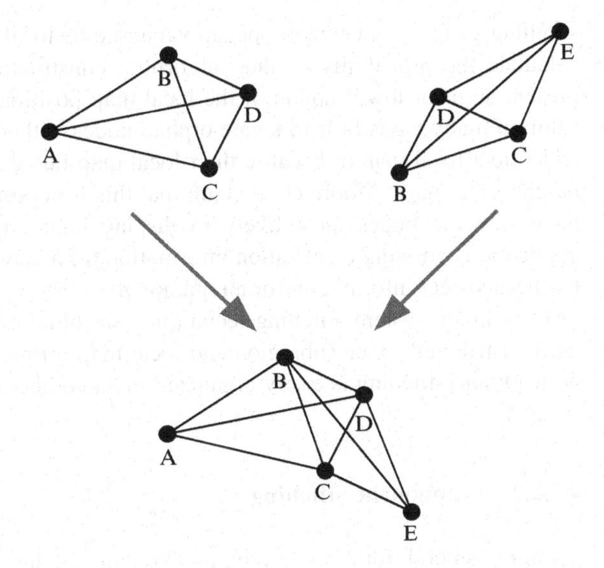

Fig. 4.4 Coordinate system stitching

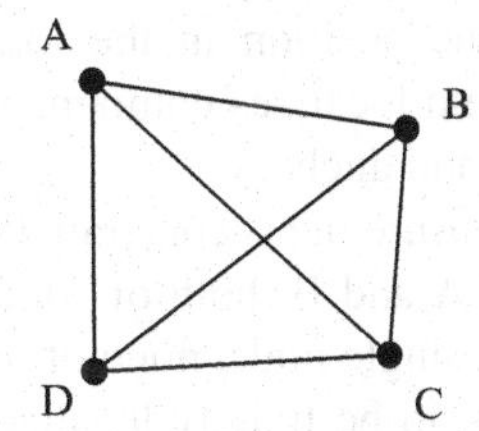

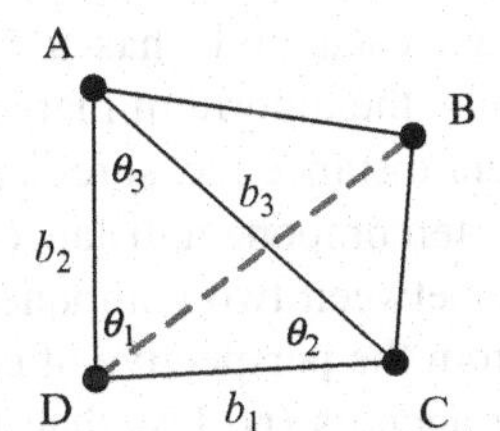

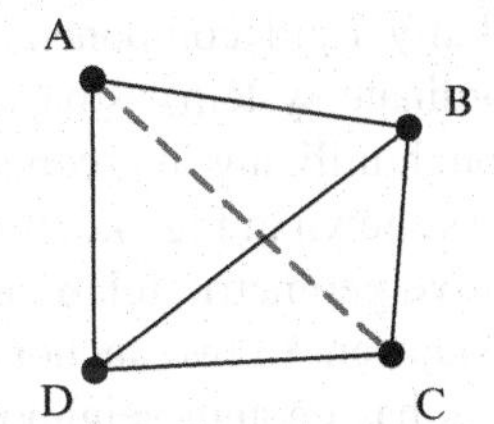

Fig. 4.5 Robust quadrilateral

Moore et al. [62] outline an approach that produces more robust local maps. Rather than using three arbitrary nodes, they use "robust quadrilateral" (robust quads) to define a local map. As shown in Fig. 4.5, a robust quad consists of four subtriangles (△ABC, △ADC, △ABD, △BCD) that satisfy

$$b \times \sin^2(\theta) > d_{\min},$$

where b is the length of the shortest side, θ is the smallest angle, and $d_{\min}$ is a predetermined parameter according to the level of measurement error. The idea is that the points of a robust quad can be placed correctly with respect to each other (i.e., without "flips"). Given zero mean Gaussian measurement error, Moore et al. demonstrate that the probability of a robust quadrilateral experiencing internal flips can be bounded by setting $d_{\min}$ appropriately. In effect, $d_{\min}$ filters out quads that have too much positional ambiguity. The appropriate level of filtering is based on the inaccuracy of distance measurements. Unfortunately, coordinate system stitching suffers from error propagation caused by local map stitching. Moore et al. calculate the probability of their algorithm constructing correct local maps and present an error lower bound of the local map positions. Furthermore, their algorithm in many cases fails to locate orphan nodes, either because they could not be added to a local map or because their local map failed to overlap sufficiently with neighboring maps. Moore et al. claim that this is acceptable because the orphaned nodes are the nodes most likely to display high error. In addition, for many applications, missing localization information for a known set of nodes is preferential to incorrect information for an unknown set.

Coordinate system stitching techniques are quite compelling. They are inherently distributed, since subregion and local map formation can trivially occur in a network and stitching is easily organized in an ad hoc manner.

4.3.2.2 Component Stitching

A more general form of coordinate system stitching is the component-based localization [65]. A component is a group of nodes that form a rigid structure (e.g., each node has finite candidate positions in the local coordinate system). Using globally rigid components (e.g., each node has a unique position in the local coordinate system) as basic units, the algorithm merges and localizes components through utilizing intercomponent distance measurements and anchors.

As shown in Fig. 4.6. three intercomponent distance measurements constrain the relative geometric relationship between two components A and B, both of which are adjacent to two anchors. From the perspective of each single node, none of the nodes has enough neighboring anchors (no less than two) to be finitely localized immediately. Traditional local-map-based algorithms will fail to localize this network, because the local maps (i.e., components A and B) do not contain anchors to convert the coordinate system. From the point of view of components, however, component A and component B can be merged into a bigger component, which is localizable by referring to the four anchors. Finally, all nodes in the two components are localized.

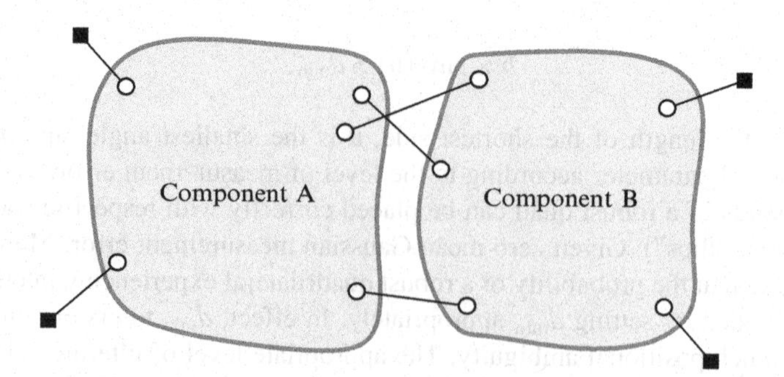

Fig. 4.6 Component-based localization

The concept of component and related terms are formally defined as follows. For a given network, a distance graph $G=(V, E)$ is constructed, where vertices denote nodes in the network and an edge (i,j) exists if nodes i,j can measure the mutual distance between them. Associated with each edge, a function $d(i,j)$: $E \rightarrow R$ is used to denote the distance value. Assume there are m anchors, labeled from 1 to m, and the left $n-m$ unknown nodes are labeled from $m+1$ to n, where n is the total number of nodes in the network. The ground truth position of node i is denoted as p_i. A realization of a network is a mapping from nodes to their 2D coordinates, P: $V \rightarrow R^2$, such that $P(i)=p_i$ for all $1 \leq i \leq m$ and $\|P(i) - P(j)\| = d(i,j)$ for all $(i,j) \in E$, where $\|P(i)-P(j)\|$ denotes the Euclidean distance of $P(i)$ and $P(j)$. Analogously, the concept of realization on the subgraph of G is defined, and the only difference is that the rotations, translations, and reflections of the mapping are treated as the same mapping when operating on a subgraph. Then, a node is localizable if and only if its image is unique for all realizations of G. A node is finitely localizable if and only if the cardinality of its image set is finite for all realizations of G. If a localization algorithm can generate a candidate position set that contains all the possible positions of a finitely localizable node, then the node is finitely localized by the algorithm. Given a distance graph G, a component is a set of nodes that has finite number of ways to be realized. A component is globally rigid if and only if there is a unique realization in a plane. Components are realized through both in-component anchors and the interconnected edges between the component and anchors. Hence, a component is realizable as long as it can determine its physical layout by the anchor information.

There are two versions of component-based localization algorithms: BCALL and CALL. As a basic version, BCALL operates on globally rigid components and unique realization of them, thus can terminate in polynomial time. BCALL has three major operations: component generation, component mergence, and component realization:

1. Component generation partitions the network into globally rigid components. Component generation follows similar procedures as generating local maps. The only difference is that a node can only join one component, such that components do not share any common node. BCALL initially selects a triangle as a component and generates the local coordinate system of the component according to the initial triangle. Other nodes then join the component and record their local coordinates through trilateration. By iterative trilateration, the newly generated component expands as large as possible. After component generation, each node either belongs to a component or becomes an isolated node.
2. After component generation, component mergence integrates nonrealizable components and isolated nodes into a larger component. As components do not share any common nodes, the mergence is performed through the interconnected edges between the two components. BCALL requires the resultant component to be globally rigid, thus there must be at least four independent interconnected edges connecting the two components. Here, independent means the edges guarantee the global rigidity of the result. For isolated nodes, it can be

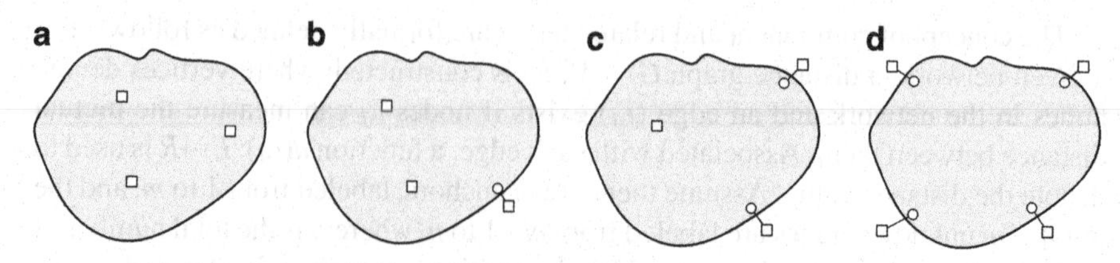

Fig. 4.7 Unique realization of components

merged to a component by trilateration. After merging, BCALL converts the local coordinate system of a component to that of the other one by solving overdetermined simultaneous equations. Component mergence is an iterative process. Some mergence can make other components or isolated nodes capable to merge into the resultant component. Component mergence process terminates when no such mergence can continue or the resultant component is realizable.

3. Component realization converts the local coordinate system of the realizable component to the physical positions. BCALL requires the realization to be unique, so the anchor information must uniquely determine the physical layout of the target component. As shown in Fig. 4.7, there are four ways to realize a component based on the number of in-component anchors:
   (a) The component contains at least three anchors
   (b) The component contains two anchors and a nonanchor node sharing an edge with a realized node
   (c) The component contains one anchor and two distinct nonanchor nodes sharing two edges with two distinct realized nodes
   (d) There are at least four independent edges connecting at least three distinct nodes in the component with at least three distinct realized nodes

CALL design is based on the concept of finite realization. CALL relaxes the requirements of the component mergence and realization from unique to finite states and adopts a consistency check step to prune the incompatible states. CALL follows similar steps as BCALL:

1. Follow the same procedure as BCALL to generate components.
2. Merge nonrealizable components and isolated nodes to generate larger components. CALL does not require the resultant component to be globally rigid. Instead, it only demands that nodes have finite candidate positions. Specifically, nodes are merged to components by bilateration. Components are merged, if the result has finite ways to be realized on a plane, thus there must be at least three interconnected edges connecting the two components.
3. Realize the components to a finite set of physical positions. Hence, the anchor information must finitely determine the physical layout of the target component. As shown in Fig. 4.8, there are three ways to finitely realize a component based on the number of in-component anchors:

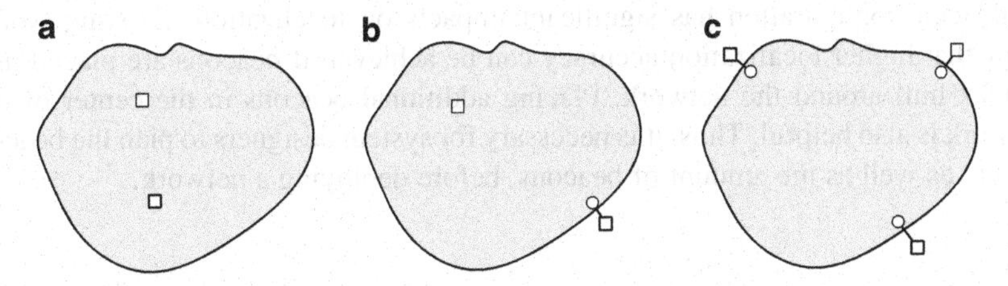

Fig. 4.8 Finite realization of components

(a) The component contains at least two anchors
(b) The component contains one anchor and a nonanchor node sharing an edge with a realized node
(c) There are at least three edges connecting the component with at least two distinct realized nodes, and there are at least two vertices associated with these edges in the component

The relaxations cause nodes to have several candidate positions. CALL inserts an extra substep in each of the above operations to check the consistency of the candidate positions. Each neighboring node pair checks the consistency of its candidate position sets by enumerating their items. Two items in each of the sets are defined as counterparts, if their distance is equal to the measured distance. A candidate position is incompatible, if it has no counterpart in the candidate position set of a neighboring node.

Using components can better share and integrate the anchor and ranging information, so the component-based localization algorithms are more applicable for sparse networks. Clearly, using fewer measurements leads the algorithm to be more sensitive for ranging errors.

4.4 Summary

We present a comparative study on existing range-based approaches with emphasis on beacon nodes, node density, accuracy, and cost.

4.4.1 Beacon Nodes

Beacon nodes (a.k.a. seeds or anchors) are necessary for localizing a network in the global coordinate system. Beacon nodes have no difference from ordinary network nodes except knowing their global locations as a priori. This knowledge can be hardcoded or acquired through some extra hardware like a GPS receiver.

Beacon configuration has significant impacts on localization. Existing works show that higher localization accuracy can be achieved if beacons are placed in a convex hull around the network. Placing additional beacons in the center of the network is also helpful. Thus, it is necessary for system designers to plan the beacon layout, as well as the amount of beacons, before deploying a network.

4.4.2 Node Density

Many localization algorithms are sensitive to node density. For instance, when the average degree is over 10, the network has a trilateration ordering with high probability, thus to suit the requirement of iterative trilateration. When designing or analyzing an algorithm, it is important to take the algorithm's implicit density assumptions into account, since high node density can sometimes be expensive and infeasible.

4.4.3 Accuracy

Given a localization algorithm, location accuracy shows how well the computed locations match with the physical positions of nodes. To be specific, location accuracy is defined as the expected Euclidean distance between the location estimate and the actual location of an unknown node, while location precision indicates the percentage of the results satisfying the predefined accuracy requirement.

Location accuracy trades off with location precision. If we relax the accuracy requirement, precision is increased and vice versa. Thus, we must put these two metrics in a common framework for comparison. We can fix location precision, say 95%, and evaluate the localization algorithms based on the corresponding accuracy performance.

Error propagation demonstrates how location accuracy varies with measurement error. Intuitively, localization error is linear with measurement error. However, it is not true for sequential localization algorithms, such as trilateration and bilateration. Nodes with large location errors would contaminate the location estimates of their neighbors. In this scenario, measurement error is no longer the only factor contributing to localization error.

4.4.4 Cost

In general, the cost of a localization system includes hardware cost and algorithm cost. Hardware cost refers to the ranging equipment. Different ranging equipments

Table 4.1 Comparative study of localization algorithms

Localization	Algorithm		Accuracy	Node Density	Beacon Amount	Cost		Error Propagation
						Comp.	Comm.	
Centralized	MDS		Median	Low	Low	High	High	Low
	SDP		High	High	Median	High	High	Low
Distributed	Beacon based	Trilateration	High	High	High	Low	Low	High[a]
		Bilateration	Low	Low	Median	High	Low	High
	Coordinate stitching	Local map	High	High	Low	Low	Median	High
		Component	Low	Low	Low	High	Median	High

[a] In case of iterative localization

provide different physical measurements, as discussed in Chap. 2. Basically, the more accurate measurements the equipments can provide, the more expensive they are. On the other hand, algorithm cost refers to the time and power consumption of computation and communication required by an algorithm. In general, distributed algorithms are more efficient than centralized ones, as they only produce local optimal solutions and exchange information locally.

After years of extensive study on this topic, many localization solutions are presented. Table 4.1 presents an overview of typical approaches in terms of accuracy, node density, beacon percentage, computation cost, communication cost, and error propagation. All approaches have their own merits and drawbacks, making them suitable for different applications. Hence, the design of a localization algorithm should sufficiently investigate application properties, as well as algorithm generality and flexibility. In present and foreseeable future study, obtaining a Pareto improvement is a major challenge. That is, increasing the performance of one of the metrics without degradation on others.

Chapter 5
Range-Free Network Localization

Range-free approaches locate nodes without the knowledge of internode distance measurements. Saving the cost of ranging hardware, they are more cost-efficient than range-based ones. Basically, range-free schemes rely on the coarse distance estimates between nodes. Hence, it is more challenging to obtain high localization.

5.1 Basic Hop-Based Algorithms

The basic idea of hop-based localization is to use hop-by-hop propagation of a network to build up node-to-anchor distance estimation. Without ranging hardware, hop counts can be used to characterize the corresponding physical distances, based on which nodes determine their locations by trilateration. According to the way of mapping hop count to distance, there are two types of hop-based schemes: DV-hop and Amorphous.

5.1.1 DV-Hop

DV-hop [59] is the basic implementation of the hop-based localization designs, which measures the internode hop counts and linearly converts them to the distance estimates by computing average per-hop distance. DV-hop follows three steps:

1. Each node estimates the least hop counts to each anchor. This could be implemented in a distributed manner by flooding a message $[x_i, y_i, h_i]$ on each anchors, where $[x_i, y_i]^T$ denotes the physical location of anchor i, h_i is a counter to record the hop counts to anchor i. The value of h_i is 1 initially and increases by 1 after each forward. Then, the received value of h_i shows the minimum hop count between the forwarding node and anchor i.
2. Anchors cooperatively estimate the per-hop distance. Once an anchor j gets the hop count h_i to anchor i, it reports the value of h_i to anchor i. After collecting

Y. Liu and Z. Yang, *Location, Localization, and Localizability: Location-awareness Technology for Wireless Networks*, DOI 10.1007/978-1-4419-7371-9_5,

these values from all other anchors, anchor i (locating at $[x_i, y_i]^T$) calculates the pre-hop distance to itself:

$$d_i = \frac{\sum_{i \neq j} \sqrt{(x_i - x_j)^2 + (y_i - y_j)^2}}{\sum_{i \neq j} h_i}.$$

3. Suppose an unknown node receives the flood messages from three anchors, say anchor i, j, and k. It uses the three distance estimates ($d_i \times h_i$, $d_j \times h_j$, and $d_k \times h_k$) to determine its location by trilateration.

5.1.2 Amorphous

It turns out that a better estimate of per-hop distance can be made if we know n_{local}, the number of neighbors per node. Suppose R is the communication range of nodes. As shown by Kleinrock and Silvester [66], it is possible to compute a better estimate for the distance covered by one radio hop:

$$d_{\text{hop}} = R\left(1 + \mathrm{e}^{-n_{\text{local}}} - \int_{-1}^{1} \mathrm{e}^{-(n_{\text{local}}/\pi)\arccos t - t\sqrt{1-t^2}} \mathrm{d}t\right).$$

We have $d_{ij} \approx h_{ij} \times d_{\text{hop}}$, where d_{ij} and h_{ij} are the distance and hop count between nodes i and j, respectively. Experimental studies [67] show that the equation above can be quite accurate when n_{local} grows greater than 5. However, when $n_{\text{local}} > 15$, d_{hop} approaches R, so the equation of d_{hop} becomes less useful. Nagpal et al. [67] demonstrate that even better hop-count distance estimates can be computed by averaging distances with neighbors. This benefit does not appear until $n_{\text{local}} \geq 15$, and, it can reduce hop-count error down to as little as $0.2R$.

5.2 Improved Hop-Based Algorithms for Anisotropic Networks

Basic hop-based algorithms assume that the network is isotropic and uniformly distributed, so that hop count represents real distance. Unfortunately, in practice, networks may be anisotropic and may contain complex inner or outer boundaries, which make the least hop counts deviating the physical distances. Let us look at the algorithms that address the anisotropy of network deployment.

5.2.1 PDM-Based Localization in Anisotropic Networks

When a network is anisotropic, hop distance (i.e., hop count) between nodes may not match physical distance well. Hence, it may introduce huge errors to use a fixed coefficient for matching hop distance to physical distance. In contrast, the match (specifically, the coefficient) may be direction and location dependent. To address this issue, Lim and Hou [68] propose an approach that builds a proximity-distance map (PDM) to represent the anisotropy of a network based on the skeleton of anchors. PDM describes the optimal linear transformations between the hop distances and the physical distances under the least-squares metric. With the help of PDM, an unknown node is able to obtain more accurate distance translation, thus to get a better location estimation.

Suppose there exist M anchor nodes. The hop distances measured from a (anchor or nonanchor) node to anchor nodes define its coordinate in a linear system. In particular, the coordinate of the ith anchor in an M-dimensional Lipschitz embedding space is represented by the proximity vector

$$p_i = [p_{i1}, \ldots, p_{iM}]^{\mathrm{T}},$$

where p_{ij} is the hop distance between the ith *anchor to the* j^{th} anchor and $p_{ii} = 0$. The overall embedding space can be represented by an $M \times M$ proximity matrix P, whose i^{th} column is the coordinate of the ith anchor:

$$P = [p_1, \ldots, p_M].$$

Here P is a square matrix with zero diagonal entries.

Similarly, the physical distance vector and matrix are defined as

$$l_i = [l_{i1}, \cdots, l_{iM}]^T \text{ and } L = [l_1, \cdots, l_M],$$

where l_{ij} is the distance of anchors i and j, which can be calculated according to the locations of the two anchors. The physical matrix L is an $M \times M$ symmetric square matrix with zero diagonal entries.

PDM is an optimal linear transformation T that gives a mapping from the proximity matrix P to the physical distance matrix L. Note that T is an $M \times M$ square matrix. Each row vector t_i of T can be obtained by minimizing the following square error:

$$e_i = \sum_{k=1}^{M} (l_{ik} - t_i p_k)^2 = ||l_i^{\mathrm{T}} - t_i P||^2.$$

The least-squares solution for the row vector t_i is

$$t_i = l_i^{\mathrm{T}} P^{\mathrm{T}} (PP^{\mathrm{T}})^{-1}.$$

As a result, PDM is defined as

$$T = LP^{\mathrm{T}} (PP^{\mathrm{T}})^{-1}.$$

The element t_{ij} of T represents the impact of the hop distance to the jth anchor node on the physical distance to the ith anchor node. Note that the main diagonal t_{ii} of T can be considered as scaling factors roughly approximating the mapping from the hop distance to the physical distance. The physical distance from a node to an anchor node is specified as a weighted sum of hop counts to all the anchor nodes. As PDM retains all the hop distance characteristics to all anchor nodes in all directions, it can precisely characterize the anisotropic relationship between proximities and physical distances.

An unknown node s can obtain its hop distance vector p_s by counting the hop counts to all anchor nodes. It then obtains the estimate of its physical distances to all anchor nodes by multiplying p_s with PDM:

$$\hat{l}_s = Tp_s.$$

PDM algorithm can be implemented in a distributed way as follows:

1. Every node initializes an empty anchor list, whose entry will be filled with the location and the hop distance for anchor nodes.
2. Every anchor node broadcasts to its neighboring nodes a probing packet containing its ID, location, and the "initial" hop distance $\{i, x_i, p_i{=}0\}$.
3. Whenever a node receives a probing packet, it calculates the new hop distance. If the new hop distance is larger than the hop distance in the anchor list, the node discards the probing packet. Otherwise, the node updates its anchor list and forwards the packet to its neighboring nodes.
4. If an anchor node b receives a probing packet containing the information for other anchor nodes, it performs step 3 as other nodes do, and updates the hop distance vector p_b. In addition, it informs the other anchor nodes of its updated p_b.
5. Whenever an anchor node b receives an update packet containing the updated p_b information, it updates both its hop distance matrix P and physical distance matrix L. After update packets from all the other anchor nodes arrive, the anchor node b computes SVD of P and obtains T.
6. A node s obtains the hop distance vector p_s from its anchor list, retrieves T from one of the anchor nodes, calculates the physical distances to anchor nodes by PDM, and estimates its location x_s by multilateration.

The main drawback of PDM is that it requires high anchor density to properly "sample" the anisotropy of the network. Moreover, it may also introduce high communicational and computational costs, considering $O(M)$ flooding messages and $O(M^3)$ cost for SVD computation, where M is the number of anchors.

5.2.2 *Rendered Path in Networks with Holes*

Trilateration is widely used for location positioning. Three nodes with known positions, often called anchors, are deployed in a network as reference points.

If nodes are able to measure their distances to the three anchors either directly or indirectly, they can calculate their positions by trilateration. Under the range-free context, however, without distance measurement, only the path information can be utilized to calculate the Euclidean distance between two nodes. The Euclidean distance represents the real geographic distance between nodes. From path information the nodes can only obtain the number of hops separating them which is denoted as hop count. As observed in [59], in isotropic networks, the hop count between two nodes can be utilized to estimate the distance between them. Thus, the distance is determined by computing the average per-hop distance multiplied by the hop count between the two nodes.

Such a design is not valid in anisotropic networks with holes. Following [69], in a homogeneous network, a hole refers to an empty area enclosed by a series of connected nodes. When a shortest path tree passes such a hole, it diverges prior to those nodes and then meets after them. Two parameters [69] σ_1 and σ_2 are defined to quantify the size of holes. Holes of considerable sizes (e.g., a percentage of the network diameter) break the isotropy of the network and may block the direct path of two nodes, curving the shortest path between them (σ_1 refers to the hop distance between the neighboring pair of nodes in two branches of the shortest path tree and their least common ancestor; σ_2 refers to the maximum hop distance between a node on one branch to the other branch). For example, as illustrated in Fig. 5.1a, when there is no hole between nodes s and t, the shortest path is close to a straight line st, and its hop number is proportional to the Euclidean distance between s and t. On the other hand, as shown in Fig. 5.1b, if there are holes, the shortest path is curved to bypass the hole. The shortest paths can actually bypass multiple holes, largely increasing the estimation error. The basic idea of rendered path (REP) is illustrated in Fig. 5.1c. After detecting the boundaries of the holes, REP labels the boundary nodes of different holes with different "colors." When a shortest path passes the holes, it is rendered with the color of the boundary nodes. A path can be rendered by multiple colors. By passing holes, a shortest path is segmented according to the intermediate "colorful" boundary nodes. The REP protocol further creates "virtual holes" to augment and render the shortest path as illustrated in Fig. 5.1c. As

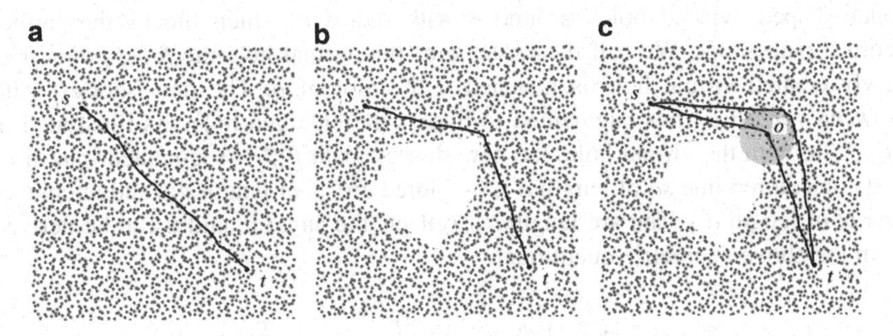

Fig. 5.1 (**a**) The shortest path between s and t is close to a *straight line*; (**b**) The shortest path between s and t is curved by a *hole* in between, (**c**) REP renders the paths and calculates the distance st from the constructed geometric structure

such, REP calculates the Euclidean distance between two nodes based on the distance and angle information along the rendered path.

Let G denote a connected region of node deployment on the plane excluding k holes inside the region. The boundary of each hole is assumed known and marked with a color C_i, $i = 1, 2, \ldots, k$, $C_i \neq C_j$ $(i \neq j)$. For any two nodes s, $t \in G$, $P_G(s,t)$ is the shortest path between them within G. Let P^G_{st} denote the Euclidean length of $P_G(s,t)$, and d_{st} denote the Euclidean distance between s and t. Clearly, $P^G_{st} \geq d_{st}$ and the objective of REP is to find d_{st} according to path information.

The REP protocol renders the shortest path $P_G(s, t)$ between s and t around intermediate holes. Every point on the boundary of a hole H is assigned with the color of H and is said to be H-colored. If there are holes in between s and t, $P_G(s,t)$, in order to be the shortest path, must intersect with the hole boundaries. From the colored points (and their colors), REP knows how many different holes the path has passed. Thus the existence of holes between two nodes can be determined from the coloring information of the shortest rendered path connecting them. The number of passed holes is equal to the number of different rendered colors.

If the path passes no holes, the length of the path $P^G_{st} = d_{st}$ and P^G_{st} can be directly used to estimate d_{st}. If the path does pass any holes, REP segments the path according to the colored points and calculates d_{st} from the length and angle information. The basic idea of REP is to create "virtual holes" around the boundary nodes on the path and augment the shortest path by forcing it to bypass those "virtual holes." REP obtains the necessary length and angle information by comparing the two shortest paths. The REP principle is shown in a basic scenario, presented in Fig. 5.2, where the shortest path between s and t intersects with a convex hole H at point o, which is H-colored, and the shortest path P^G_{st} is segmented into so and ot. Assume that $\|so\| = d_1$ and $\|ot\| = d_2$. As Fig. 5.2a shows, according to *law of cosines*, there exists the following mathematical relationship in the triangle Δsot: $\|st\|^2 = \|so\|^2 + \|ot\|^2 - 2\|so\|\|ot\|\cos\angle sot$. Thus,

$$d_{st} = \sqrt{d_1^2 + d_2^2 - 2d_1 d_2 \cos\alpha}.$$

To obtain the angle quantity α between so and ot, REP creates an approximately round-shaped "virtual hole" around o with radius r, which blocks the former shortest path s–o–t. We call the center o of this virtual hole the *focal point*. The newly created virtual hole is attached with color of o. The new shortest path between s and t is thus augmented to bypass the enlarged hole. As illustrated in Fig. 5.2b, with the virtual hole, the new shortest path P^{G*}_{st} is segmented into three parts: uncolored line sa of length d'_1, o-colored arc $\widehat{ab}$ of length d_{ab} and uncolored line bt of length d'_2. The arc length d_{ab} reflects the angle θ, and α can be derived from the above geometric quantities:

$$\alpha = 2\pi - \frac{d_{ab}}{r} - \arccos\frac{r}{d_1} - \arccos\frac{r}{d_2}.$$

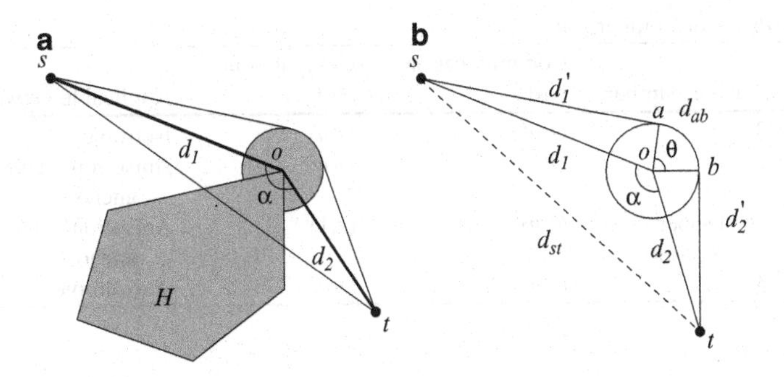

Fig. 5.2 A basic scenario for REP

Using these formulas, the Euclidean distance d_{st} can be calculated from the length information in the two rendered paths P_{st}^{G} and P_{st}^{G*}. Thus, if the shortest path P_{st}^{G} between two points s and t intersects with some hole at a single point o, d_{st} is computable by augmenting the shortest path and using the length information in rendered paths. Indeed, the basic principle can be generalized to deal with many more complicated cases, such as the convex hole, multiple holes, or even concave holes.

Thus far, we have described the principle of the REP protocol in the continuous domain. In a real deployed sensor network, however, sensors are distributed discretely on the field. Also, due to the lack of global coordination, the methods of coloring the nodes, rendering the paths, and disseminating the coloring information in a distributed manner need to be carefully addressed. The practical REP protocol includes five major components: *boundary detection*, *shortest path exploration*, *virtual hole construction*, *virtual shortest path construction*, and *distance computing*. The protocol proceeds as follows. First, the system detects and enumerates the holes inside the hole boundary as well as the nodes on the boundary using the algorithm in [69]. Then, each node explores the shortest path to the three anchors and calculates the Euclidean distances to them by rendering and augmenting the shortest paths. Based on the estimated distances to the anchors, the nodes localize themselves by triangulation. All operations are carried out in a distributed fashion among discrete sensor nodes.

We summarize REP protocol in several aspects including protocol features, applicability, and overhead. Under the range-free context, REP can utilize as few as three anchors to localize nodes in anisotropic networks. REP does not presume super anchors. Each anchor is assumed to have the same communication capability as an ordinary node. To calculate the location, each node needs several rounds of query flooding to find different rendered paths and accordingly calculate the Euclidean distances to the anchors. With the help of hole combination and parallel path construction the rounds of query flooding are limited within a constant <9 for a single node to all three anchors. Consequently for an entire network, the communication overhead is bounded by $O(n^2)$ where n is the number of nodes in the network.

Table 5.1 Protocol Comparison

Protocol	Anchor number	Communication cost	Computation cost	Applicable networks
DV-hop	3	$O(n^2)$	$O(n)$	Isotropic
PDM	$O(n)$	$O(n^2)$	$O(n^3)$	Anisotropic uniform anchors
APIT	$O(n)$ super anchors	$O(n)$	$O(1)$	Anisotropic uniform anchors
REP	3	$O(n^2)$	$O(nL)$	Anisotropic

The anchor nodes bear most of the computational burden. Each anchor deals with distance queries from all the network and for each query the anchor does at most $O(L)$ computations to calculate the distance from the rendered path, where L is the number of holes within the network. Thus for each anchor, the computation overhead is $O(nL)$.

Table 5.1 compares REP with the three state-of-the-art range-free approaches: DV-hop [59], PDM [68], and APIT [41]. DV-hop presumes isotropic networks and triangulates the node location with its network distances to the three anchors. Each node floods the network for computing the hop counts so the communication cost of DV-hop is $O(n^2)$. Each anchor accepts requests from all the network and sends out feedback with $O(n)$ computation cost. PDM is a space-embedding approach which with the help of a portion of anchors can handle anisotropic networks. Relying on each node flooding the network to estimate the hop counts to all the anchors, PDM has $O(n^2)$ communication cost. For each anchor, the cost to compute the transformation matrix is $O(n^3)$. APIT is a typical connectivity-based approach employing super anchors with much larger transmitting radii than ordinary nodes. The anchors locally broadcast their locations and the undetermined nodes do not send any requests. They only listen to the anchors and determine their location locally. Thus the communication cost is $O(n)$ and the computation cost is $O(1)$.

5.2.3 *Delaunay-Complex-Based Localization*

A major challenge in range-free localization is flip ambiguity. That is, two triangles sharing an edge can be embedded in two possible ways, with the two triangles on the same side or on opposite sides of the common edge. As shown in Fig. 5.3, preserving all internode distances, there are two possible ways to embed the network. In the view of a whole network, a more serious case can happen as a global flip, in which a part of the network may have two ways to be embedded reflecting a line.

To address this problem, Delaunay-complex-based algorithms are introduced [70, 71]. The Delaunay complex is defined in the notion of abstract simplicial complex [71]. An example of Delaunay complex is shown in Fig. 5.4c, where the shaded area denotes four cocircular landmarks and this corresponds to a simplex of

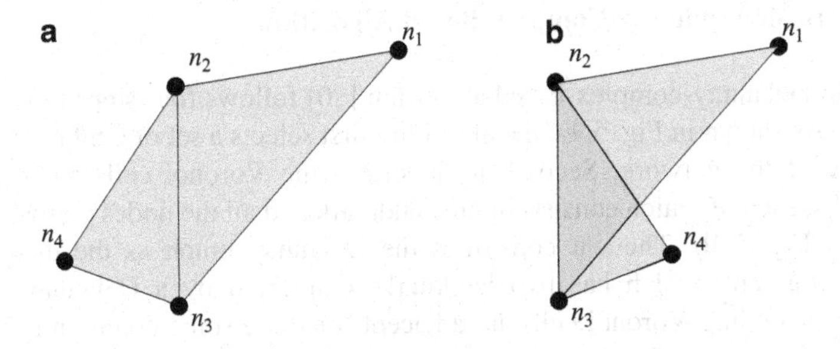

Fig. 5.3 Flip ambiguity

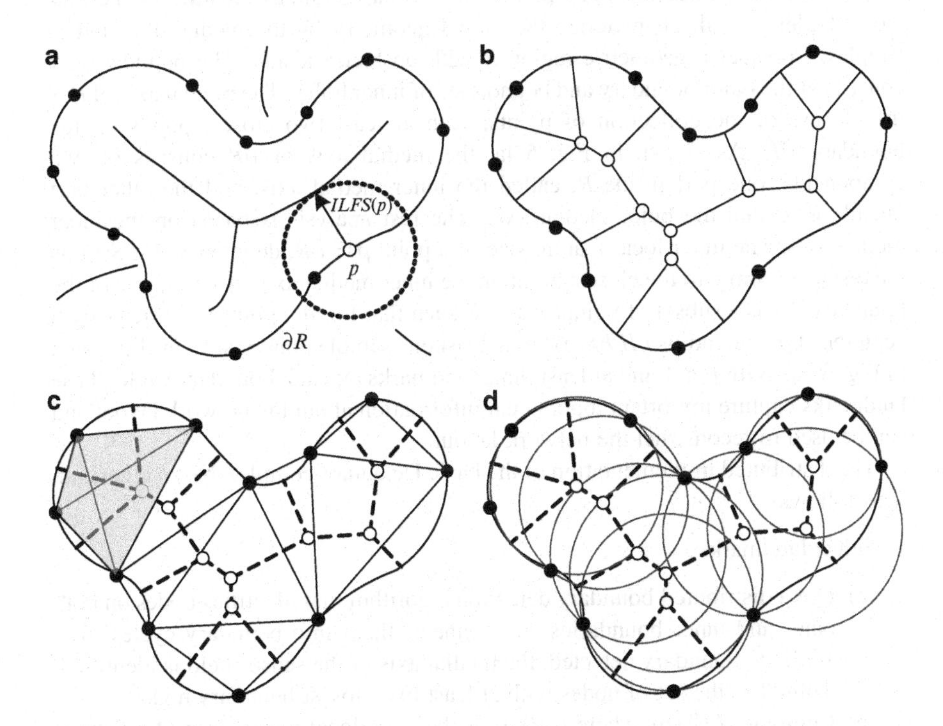

Fig. 5.4 Landmark selection scheme

dimension 3. An important characteristic of Delaunay complex is that the Delaunay complexes do not share any common nodes except the boundary. For example, in Fig. 5.3, if $n_1n_2n_3$ and $n_2n_3n_4$ are Delaunay complexes, the embedding shown in Fig. 5.3b is not legal, because there are nodes lying in both the complexes. Hence, the embedding of combinatorial Delaunay complex is unique with sufficient landmarks, where landmarks are the points to generate Delaunay complex.

5.2.3.1 Basic Delaunay-Complex-Based Algorithm

The basic Delaunay-complex-based algorithm [70] follows four steps to localize a network. As shown in Fig. 5.4a, the algorithm first selects a set of landmarks on the boundary of the network. Second, it constructs the Voronoi cells based on the landmarks, each of which consists of one landmark and all the nodes closest to it, as shown in Fig. 5.4b. Then, it constructs the Delaunay graph as the dual of the Voronoi diagram, which has two landmarks connected by a Delaunay edge if their corresponding Voronoi cells are adjacent (or share some common nodes), as shown in Fig. 5.4c. Third, the complexes are embedded in a plan, forming a skeleton of the network global layout. Finally, the nonlandmark nodes can be located by the measurements to the nearby landmarks.

The landmark selection must present the global geometric feature of the network. Lederer et al. characterize the global geometry by the medial axis and r-simples. Consider a geometric region R with obstacles inside. The boundary ∂R consists of the outer boundary and boundaries of inner holes. The medial axis of R is the closure of the collection of points, with at least two closest points on the boundary ∂R. As shown in Fig. 5.4a, the medial axis of ∂R consists of two components, one part inside R, called the inner medial axis, and the other part outside R, called the outer medial axis. The rest analysis is based on the inner medial axis. The inner local feature size of a point $p \in \partial R$, denoted as $ILFS(p)$, is the distance from p to the closest point on the inner medial axis. An r-sample of the boundary ∂R is a subset of points S on ∂R such that for any point $p \in \partial R$, the ball centered at p with radius $r\, ILFS(p)$ has at least one sample point inside, as illustrated in Fig. 5.4a. With $r < 1$ and at least three landmarks on each boundary cycle, these landmarks capture important topological information about the network layout and can be used to reconstruct the network layout.

The distributed implementation of the basic Delaunay-complex-based algorithm is as follows:

1. Select landmarks
   (a) Use a distributed boundary detection algorithm that identifies nodes on both outer and inner boundaries and connects them into boundary cycles [69]. With the boundary detected, the medial axis of the sensor field is identified, defined as the set of nodes with at least two closest boundary nodes.
   (b) Compute $ILFS(p)$, where $ILFS(p)$ is the inner local feature size of p defined as the hop count distance from p to its closest node on the inner medial axis. Each node p obtains $ILFS(p)$ by recording the minimum hop count from the messages broadcasted by the nodes on the medial axis.
   (c) Select landmarks from boundary nodes such that for any node p on the boundary, there is a landmark within distance $ILFS(p)$. A serial way is to use a message traversing along the boundary cycles and select landmarks along the way in a greedy fashion to guarantee the sampling criterion. Alternatively, to achieve lower time cost, let each boundary node p wait for a random period of time and select itself as a landmark. Then p sends a

suppression message with TTL as *ILFS*(p) to adjacent boundary nodes. A boundary node receiving this suppression message will not further select itself as landmarks.

2. Compute Voronoi diagram and combinatorial Delaunay complex
   (a) Construct Voronoi cells. All the landmarks flood the network simultaneously and each node records the closest landmark(s). A node p will not forward the message if it carries a hop count larger than the closest hop count p has seen. So the propagation of messages from a landmark l is confined within l's Voronoi cell.
   (b) Construct Delaunay complex. A k-witness node w, defined as a border node which is within 1-hop from interior nodes of k different Voronoi cells, reports to the corresponding k landmarks. Such a report contains the IDs of the landmarks involved in this dimension $k-1$ Delaunay simplex, together with the distance vector from the witness node w to each of the k landmarks.

3. Embed Delaunay complex
   (a) Initialize a coordinate system. Embed one simplex S_1 arbitrarily.
   (b) Embed other Delaunay complexes. For a neighboring simplex S_2, let l_1 and l_2 be the landmarks they share in common. For each landmark l_i in S_2 not yet embedded, compute the two points that are with distance $d(l_1; l_i)$ from l_1 and $d(l_2; l_i)$ from l_2, where $d(;)$ is the hop-count distance between landmarks, estimated in the previous section. Among the two possible locations we take the one such that the orientation of points $\{l_1, l_2, l_i\}$ is different from the orientation of $\{l_1, l_2, l_r\}$, where l_r is any landmark of S_1, other than l_1 and l_2. Thus l_i and l_r lie on opposite sides of edge l_1l_2.

4. Network localization
   (a) Since the locations of the landmarks are known, each nonlandmark node just runs a trilateration algorithm to find its location by using the hop count estimation to the landmarks.

5.2.3.2 Incremental Landmark Selection Scheme

A drawback of the basic Delaunay-complex-based algorithm is the requirement of boundary detection. Boundary detection leads to high computational and communicational costs in a large-scale network, and may be infeasible in sparse networks. To address this issue, Wang et al. propose an incremental landmark selection scheme that does not rely on the knowledge of boundaries [71].

The main idea is to use the union of Voronoi balls to approximate the region R. Voronoi ball is formally defined as follow. A point is called a Voronoi vertex if it has equal distance to at least three landmarks. The Voronoi vertices inside R are called the inner Voronoi vertices. A ball $B_r(p)$ centered at an inner Voronoi vertex p with radius r

equivalent to the distance from p to the closest landmarks is called a Voronoi ball. As shown in Fig. 5.4d, the union of the Voronoi balls approximately covers the whole network. The uniqueness of embedding and good coverage of the induced Delaunay complex is guaranteed by the following local conditions of landmark selection:

1. *Local Voronoi edge connectivity*. The Voronoi edges for each landmark u form a connected set.
2. *Local Voronoi ball coverage*. Each node x inside a Voronoi cell $V(u)$ is δ-covered by a Voronoi ball $B_r(p)$, where p is a Voronoi vertex with landmark u. The Delaunay complex DC(L) δ-covers R if every point $x \in R$ will be within distance $(1+\delta)$r from the center p of a Voronoi ball $B_r(p)$, where r is the radius of this Voronoi ball.

By the two conditions, the selected landmarks always locate at the boundary of the network, hence this landmark selection scheme avoids boundary detection.

The distributed implementation of the landmark selection algorithm is as follows:

1. *Two initial landmarks are selected on the boundary*. In order to guarantee that these two starting landmarks are on the boundary, a message is flooded by a random node r to find the farthest node p from r, and p must be on the network boundary. Then, another message is flooded by p to find the farthest node q from p. Node q will be on the boundary as well. Nodes p and q are two initial landmarks.
2. *Compute Voronoi diagram*. Each landmark learns of its closest landmark(s) and all the nodes with the same closest landmark are naturally classified to be in the same Voronoi cell. Nodes with more than one closest landmarks lie on a Voronoi edge or vertex. Voronoi vertex is a node with equal distance to at least three landmarks.
3. *Select more landmarks incrementally*. With the Voronoi diagram from the initial two landmarks, more landmarks are selected incrementally. For each landmark u and its Voronoi cell $V(u)$, do the following checks:
   (a) If the Voronoi edges of u are not connected (this can be checked by having each connected component of the union of u's Voronoi edges sending a message to u), choose among all nodes that are endpoints of Voronoi edges lying on the network boundary and select the one furthest from u as a new landmark.
   (b) If the Voronoi edges of u are connected, check each point p in Voronoi cell $V(u)$ and any Voronoi vertex v associated with u. Point p is selected as a new landmark if p is furthest away from any relaxed Voronoi ball $B_{(1+\delta)r}(v)$ among all points that are not yet δ-covered by Voronoi balls of u. Here r is the hop-count distance between u and v.

Delaunay-complex-based algorithms can properly reconstruct the global geometry of the network by purely connectivity information. However, the computational and communicational cost of this design depends on the amount of landmarks, i.e., the complexity of the network layout, which is not stable for variously configured networks.

5.3 Proximity-Based Algorithms

RSS is closely related to the proximity of neighboring nodes. However, directly using RSS for physical distance estimation is unacceptable in many scenarios because of unknown radio path loss factors, multipath effects, hardware discrepancies, antenna orientation, etc. Thus, a number of localization approaches try to explore the closeness information according to RSS measurements instead of the ranging information.

5.3.1 Point-in-Triangulation Test

Approximate point-in-triangulation test (APIT) is an area-based localization algorithm leveraging the underlying proximity estimation of RSS [41]. APIT algorithm follows three main steps. First, APIT estimates the possible area for a target node by testing whether the node is inside the triangles formed by three nearby anchors. Second, APIT aggregates the results of the area estimation by intersecting all possible areas. Third, APIT takes the center of the intersection as estimated position of the target node.

The basic block for APIT is called the point-in-triangulation test (PIT). PIT is based on the following theorem: given a point M and a triangle ΔABC, if there exists a direction such that a point adjacent to M is further/closer to points A, B, and C simultaneously, then M is outside of ΔABC; otherwise, M is inside ΔABC, as illustrated in Fig. 5.5. In APIT, an unknown node chooses three anchors from all audible anchors and tests whether it is inside the triangle formed by connecting these three anchors. As the target node cannot move to perform the ideal PIT test, APIT adopts an approximation in the discrete domain based on the assumption of moderate network density (with connectivity above 6). That is, if no neighbor of a target node is further from/closer to all three anchors A, B, and C simultaneously, M is assumed to be inside triangle ΔABC; otherwise, M is assumed to reside outside this triangle. The way to compare the distances is based on the RSS of each node on anchors. APIT assumes that the received signal strength is monotonically decreasing in an environment without obstacles in a certain propagation direction. Hence, further nodes introduce lower RSS on anchors.

APITs aggregate the result of PITs through a grid array representing the maximum area in which the target node will likely reside. For each APIT inside decision (a decision where the PIT test determines the node is inside a particular region), the values of the grid regions over which the corresponding triangle resides are incremented. For an outside decision, the grid area is similarly

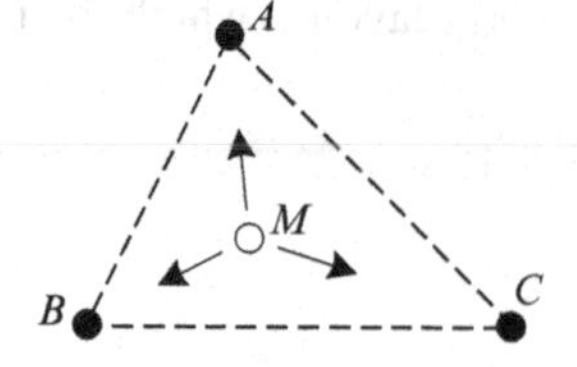

Fig. 5.5 Point-in-triangulation test

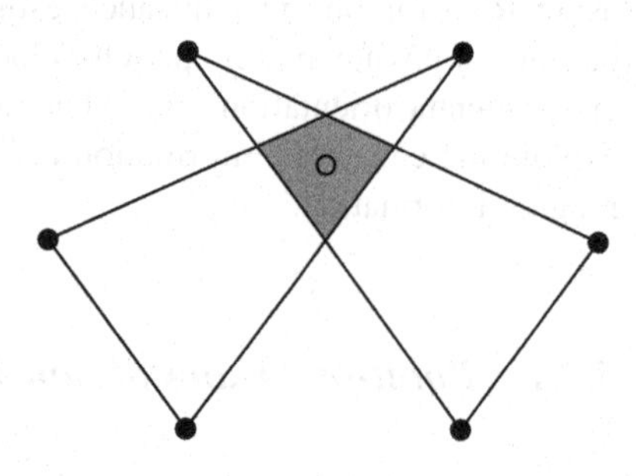

Fig. 5.6 Aggregation of the PIT results

decremented. Finally, the location of the target node is estimated as the center of maximum overlapping area (i.e., the grids with maximum value), as illustrated in Fig. 5.6.

PIT test only requires the monotonicity of the RSS, which is more practical and realistic in practice. Hence, APIT can generate relatively stable result based on the inconstant RSS. However, the estimated area size of APIT is determined by the density of triangles. As a result, APIT requires large proportion of anchors to achieve high accuracy.

5.3.2 Perpendicular Intersection

Errors are often inevitable and unpredictable for RSS-based ranging techniques. As an early attempt to address this issue, perpendicular intersection (PI)-based localization [72] uses a mobile beacon to explore the mapping between RSS and distance, as described in Fig. 5.7. The mobile beacon continuously broadcasts signals to the rest static nodes and moves from 10 m away from O to 20, 30, 40, and 50 m. The rest nodes are placed on the line perpendicular to the trajectory OA. All the measured RSS values are shown in Fig. 5.8 corresponding to the node deployment in Fig. 5.7. The results reveal that the closer a node to the signal sender (node A), the larger RSSI value it perceives. Based on such observation, PI utilizes the geometric relationship of perpendicular intersection and computes node positions by contrasting RSS values measured at each static node.

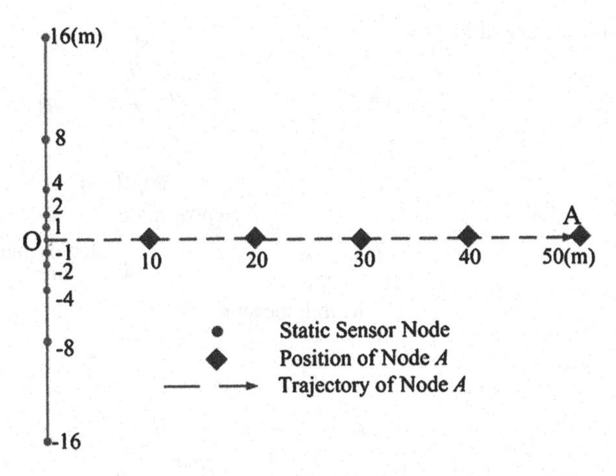

Fig. 5.7 Deployment sketch of the observational experiment

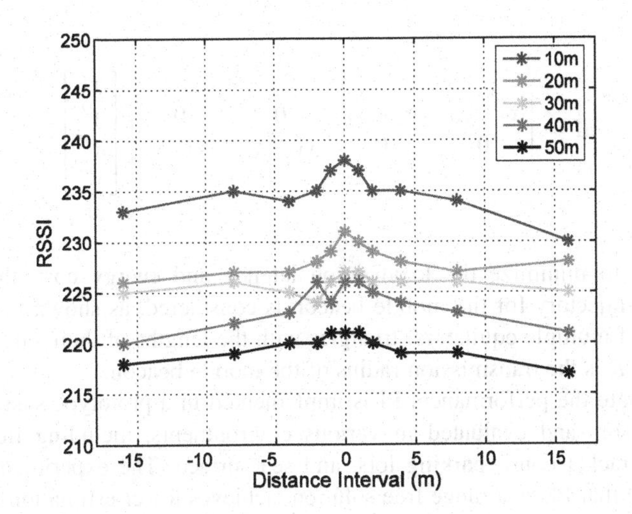

Fig. 5.8 The observed RSSI values

The basic scheme of PI is illustrated using Fig. 5.9. When the mobile beacon moves along a straight line, the largest RSS value received by a sensor node N often, if not always, corresponds to the point on the line that is closest to the node. Theoretically, this point should be the projection of the node on the line. When the mobile beacon moves along two different lines, e.g., P_1P_2 and P_2P_3, there will be two different projections of the node on the trajectory, i.e., A and B. Thus node N can be located as the intersection point of two perpendiculars (AN and BN) that cross the mobile beacon's trajectory over the two projections, respectively. The coordinates of N are calculated using the following equation:

Fig. 5.9 The basic scheme of PI

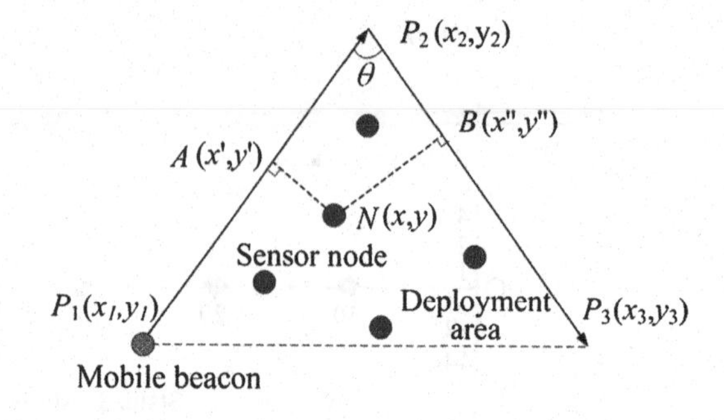

$$\begin{bmatrix} x \\ y \end{bmatrix} = \begin{bmatrix} x_2 - x_1 & y_2 - y_1 \\ x_3 - x_2 & y_3 - y_2 \end{bmatrix}^{-1} M,$$

where

$$M = \begin{bmatrix} x_2 - x_1 & y_2 - y_1 & 0 & 0 \\ 0 & 0 & x_3 - x_2 & y_3 - y_2 \end{bmatrix} \begin{bmatrix} x' \\ y' \\ x'' \\ y'' \end{bmatrix}.$$

In order to minimize the localization latency and energy cost, the optimal movement trajectory for the mobile beacon is considered, as shown in Fig. 5.10. It consists of multiple equilateral triangles with the lengths of their sides all equal to R, where R is the transmission radius of the mobile beacon.

To evaluate the performance, PI is implemented in a prototype system of 100 TelosB sensors and evaluated in various environments, including library hall, laboratory, racket court, parking lots, and sea surface. The experimental results demonstrate that PI, as a range free solution, achieves lower estimation errors and more stable precisions.

5.4 Relative Distance Estimation

Regulated signature distance (RSD) [73] is designed to measure the proximity by RSS without the knowledge of radio attenuation model. RSD is based on the following observation: in outdoor open-air scenarios, the radio signal strength weakens approximately monotonically with the physical distance, especially from the viewpoint of a single node. They conduct a real-world experiment to testify this

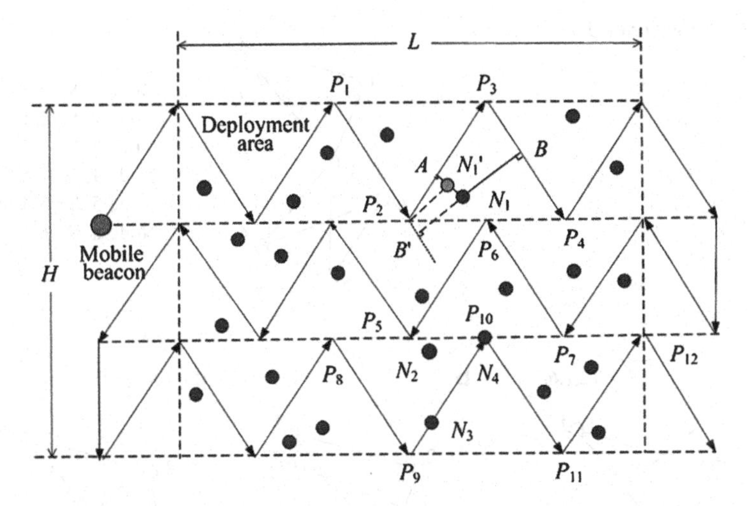

Fig. 5.10 A sensor network and the optimal trajectory of the mobile beacon

assumption, and the result shows that this proposition holds for about 88% cases in average. RSD computation follows three steps: (1) neighborhood ordering by RSS; (2) calculating signature distance (SD); (3) SD regulation.

Given the RSS sensing results for neighboring nodes, a node can obtain a neighborhood ordering with two steps:

1. Sorting its 1-hop neighbors according to their signal strength by decreasing order
2. Adding itself as the first element in the sorted node list

Figure 5.11 shows the result of the ordering of a network. The ordering is defined as a signature of the node.

The SD of two nodes is calculated by the flips of their signatures. Given two signatures S_i and S_j, a flip of the signature pair is that the ordering of nodes u_m and u_n in S_i gets reversed in S_j. For example, the ordered node pair {1,6} in $S_2 = (2,1,6,3)$ gets reversed to {6,1} in $S_5 = (5,4,6,1)$. There are three types of flips: explicit flip, implicit flip, and possible flip. If node u_m and u_n appear in both S_i and S_j and get reversed order, it is an explicit flip, e.g., node pair {1,6} in S_2 and S_5. Implicit flip is related to the node that does not exist in one of the signatures, for example node 2 in S_2 and S_5. As this implies that node 2 and node 5 cannot communicate with each other, node 2 is further than all nodes in S_5 in the view of proximity. Hence, for each signature, there implicitly exists a wildcard, denoted by □, in the end of the list that matches any node not in the signature. An implicit flip is a flip by using an implicit wildcard in a signature, e.g., node pair {2,6} in $S_2 = (2,1,6,3,□)$ and $S_5 = (5,4,6,1,□)$. Possible flip is related to a pair of nodes that do not exist in one of the signatures, for example node pair {2,3} in S_2 and S_5. As the orders (2,3)

Fig. 5.11 Neighborhood ordering

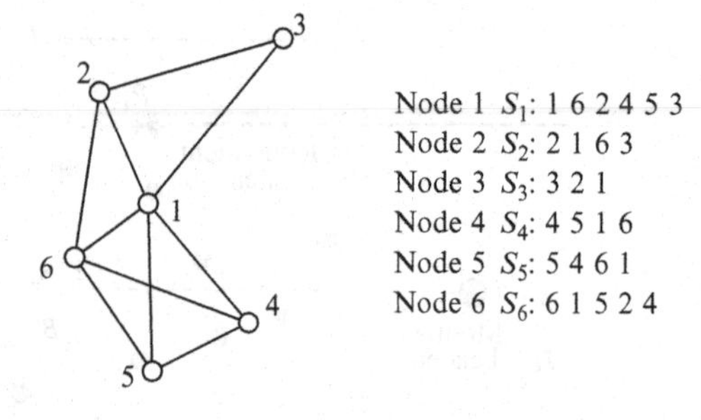

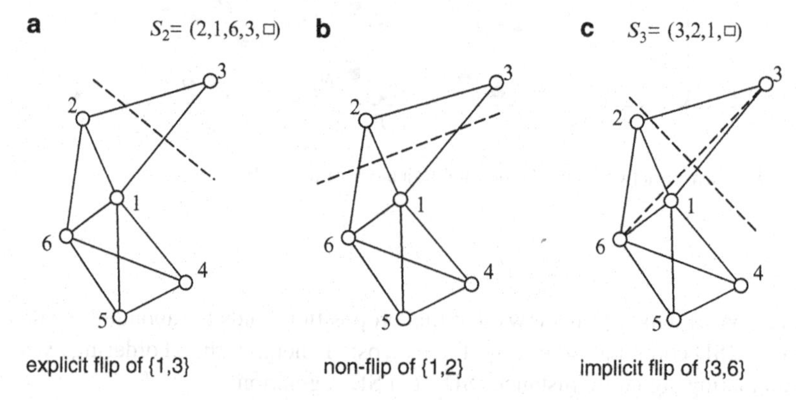

Fig. 5.12 Physical meaning of flips: (**a**) explicit flip of {1,3}; (**b**) nonflip of {1,2}; and (**c**) implicit flip of {3,6}

and (3,2) are both legal for matching the wildcards, this gives a possible node-pair flip with 50% probability. In a word, the signature distance SD(S_i,S_j) is equal to the summation of the number of explicit flips $F_e(S_i,S_j)$, implicit flips $F_i(S_i,S_j)$, and possible flips $F_p(S_i,S_j)$ times 0.5 (50% probability of flip for possible node pairs), namely, $\mathrm{SD}(S_i,S_j) = F_e(S_i,S_j)+F_i(S_i,S_j)+F_p(S_i,S_j)\times 0.5$.

The number of flips between nodes is related to the mutual physical distance. Formally, a node-pair flip $(u_m, u_n)\Rightarrow(u_n, u_m)$ from S_i to S_j indicates that the line segment $L(u_i,u_j)$ passes the perpendicular bisector line $B(u_m, u_n)$. As illustrated in Fig. 5.12, the ordered node pair (u_m, u_n) in S_i means that from node i's point of view, node u_m is closer than node u_n. In other words, if we divide the plane with $B(u_m, u_n)$, the different ordering of (u_m, u_n) in S_i and S_j indicates that node S_i and node S_j are located on the different side of $B(u_m, u_n)$. Based on the definition of signature distance, SD(S_i,S_j) evaluates the difference between two signatures S_i and S_j by counting the total number of node-pair flips. Therefore, SD(S_i,S_j) is equivalent to the number of bisector lines we need to pass if going from neighboring node u_i to u_j along the line segment $L(u_i,u_j)$.

Spatially nonuniform bisector line density could affect the effectiveness of SD as a relative distance. The problem comes from two aspects: (1) local node placement

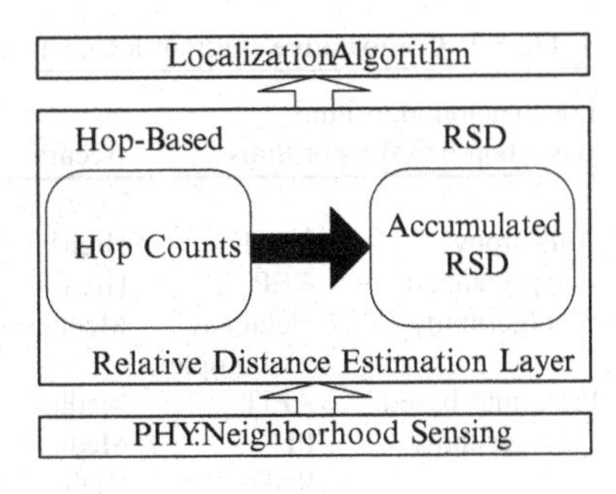

Fig. 5.13 RSD embedding

and (2) network-wide neighborhood size. Hence, the original SD is regulated by the density and neighborhood size, given by

$$\mathrm{RSD}(u_i,u_j) = \mathrm{SD}(S_i,S_j)\frac{\sqrt{K}}{K(K-1)/2},$$

where $K = \|S_i \cap S_j\|$ is the total number of nodes in the neighborhood of node u_i and u_j combined. In this equation, $K(K-1)/2$ calculates the number of local bisector lines, used to normalize $\mathrm{SD}(S_i,S_j)$ with the local bisector density; $\sqrt{K}$ estimates the diameter of this neighborhood, which puts the factor of neighborhood size into consideration.

The design of RSD can be implemented in a supporting layer that is transparent to the localization algorithms. As illustrated in Fig. 5.13, RSD is used to estimate the distance between two nodes instead of the shortest-path hop count. Specially, the accumulated RSD between two nodes is defined as

- For 1-hop neighboring nodes u_i and u_j , accumulated RSD equals $\mathrm{RSD}(u_i,u_j)$
- For nonneighboring nodes u_i and u_j, accumulated RSD is calculated as the summation of the RSD values of neighboring nodes along a path between u_i and u_j

RSD obtains higher accuracy than hop-based measurement without the calibration of model or environment profiling. However, this design relies on assumption that the monotonicity of RSS holds over the whole network, which may not be always practical in complex environments.

5.5 Summary

Table 5.2 provides an overview of the range-free localization approaches in terms of accuracy, node density, beacon percentage, computation cost, communication cost, and error propagation. All approaches have their own advantages and requirements, making them suitable for different applications.

Table 5.2 Comparative study of localization algorithms

Localization algorithm			Node	Beacon	Cost		Error
Basic hop-based algorithms		Accuracy	density	amount	Comp.	Comm.	propagation
		Low	Low	Low	Low	Low	Low
Anisotropy-specialized algorithms	PDM	High[a]	Median	Median	High	High	Low
	REP	High	Median	Low	Low	High	Low
	Delaunay-complex	Median	Median	Low	Low	High	Median
Proximity-based algorithms	APIT	Median	High	High	High	Low	Low
	PI	Median	Low	Low	Low	Low	Low
	RSD	High	High	Low	Median	Median	Low

[a] Depends on the implementation

Chapter 6
Error Control

Many Localization algorithms are range based and adopt distance ranging techniques, in which measuring errors are inevitable. Generally speaking, errors fall into two categories: *extrinsic* and *intrinsic*. The extrinsic error is attributed to the physical effects on the measurement channel, such as the presence of obstacles, multipath and shadowing effects, and the variability of the signal propagation speed, due to changes in the surrounding environment. On the other hand, the intrinsic error is caused by limitations of hardware and software. While extrinsic error is more unpredictable and challenging to handle in realistic deployments, the intrinsic one can also cause many complications when using multihop measurement information to estimate node location. Results from field experiments demonstrate that even relatively small measurement errors can significantly amplify the error in location estimates [62]; thus, for high-accuracy localization algorithms, error control is essential.

6.1 Measurement Errors

6.1.1 Errors in Distance Measurements

Table 6.1 lists the typical measuring (intrinsic) error of a range of nowadays ranging techniques: TDoA, RSS in AHLoS [31], ultra-wideband system [74], RF time-of-flight (ToF) ranging systems [75], and elapsed time between the two time of arrival (EToA) in BeepBeep [34]. In general, the accuracy of RF-based ranging techniques, e.g., RSS, UWB, and RF ToF, can achieve the meter-level accuracy in a range of tens of meters. In contrast, ToA-based methods have more accurate results in the order of centimeters but require extra hardware and energy consumption.

On the other hand, extrinsic errors are caused by environmental factors or unexpected hardware malfunction, leaving difficulties on characterizing them. We will review the state-of-the-art works on controlling the intrinsic and extrinsic errors in the following sections of location refinement and outlier-resistant localization, respectively.

Y. Liu and Z. Yang, *Location, Localization, and Localizability: Location-awareness Technology for Wireless Networks*, DOI 10.1007/978-1-4419-7371-9_6,

6.1.2 Negative Impact of Noisy Ranging Results

Errors in distance ranging make localization more challenging in the following four aspects [76]:

1. ***Uncertainty***. Figure 6.1 illustrates an example of trilateration under noisy ranging measurements. Trilateration often meets the situation that the three circles do not intersect at a common point. In other words, there does not exist any position satisfying all distance constraints.
2. ***Nonconsistency***. In many cases, one node has many reference neighbors. Any subgroup of them (on less than three) can locate this node by multilateration. The computed results, however, is varying if different groups of references are chosen, resulting in nonconsistency. Thus, when alterative references are available, it is a problem to determine which combination of references provides the best result.
3. ***Ambiguity***. The existence of flip and flex [62, 77] may lead to large localization errors. Although localizability theory presents methods of detecting possible flip and flex ambiguities, these methods do not work when distance measurements are noisy.

Table 6.1 Measurement accuracy of different ranging techniques

Technology	System	Measurement accuracy	Range
TDoA	AHLoS	2 cm	3-10 m
RSS	AHLoS	2-4 m	30-100 m
UWB	PAL UWB	1.5 m	N/A
RF ToF	RF ToF ranging system	1-3 m	100 m
EToA	BeepBeep	1-2 cm	10 m

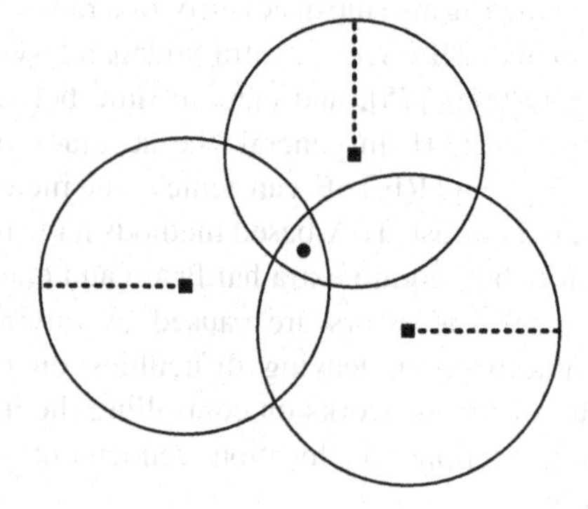

Fig. 6.1 Trilateration under noisy ranging measurements

4. ***Error propagation***. The results of a multihop localization process are based on a series of single hop multilaterations in an iterative manner [31]. In such a process, errors, coming from each step of multilateration, propagate and accumulate [60, 61].

6.2 Error Characteristics

Localization error is a function of a wide range of network configuration parameters, including the numbers of beacons and to-be-localized nodes, node geometry, network connectivity, etc., which constitute a complicated system. Understanding the error characteristics is an essential step toward controlling error. The Cramer Rao lower bound (CRLB) provides a means for computing a lower bound on the covariance of any unbiased location estimate that uses RSS, TdoA, and other ranging techniques. In addition, CRLB can serve as a benchmark for a particular localization algorithm. If the bound is closely achieved, there is little gain to continue improving the algorithm's accuracy. Furthermore, the dependence of CRLB on network parameters helps to understand the error characteristics of network localization.

6.2.1 What is CRLB

The Cramer Rao Lower Bound (CRLB) is a classic result from statistics that gives a lower bound on the error covariance for an unbiased estimate of parameter [78]. This bound provides a useful guideline to evaluate various estimators. One important and surprising advantage of CRLB is that we can calculate the lower bound without even considering any particular estimation method. The only thing needed is the statistical model of the random observations, i.e., $f(X|\theta)$, where X is the random observation and θ is the parameter to be estimated. Any unbiased estimator $\hat{\theta}$ must satisfy

$$\mathrm{Cov}(\hat{\theta}) \geq \{-E[\nabla_\theta(\nabla_\theta \ln f(X|\theta))^{\mathrm{T}}]\}^{-1}, \tag{6.1}$$

where $Cov(\hat{\theta})$ is the error covariance of the estimator, $E[\cdot]$ indicates expected value, and ∇_θ is the gradient operator with respect to θ.

The CRLB is limited to unbiased estimators, which provides estimates that are equal to the ground truth if averaged over enough realizations. In some cases, however, a biased estimation approach can produce both a variance and a mean-squared error that are below the CRLB.

6.2.2 CRLB for Multihop Localization

In network localization, the parameter vector θ of interest consists of the coordinates of nodes to be localized, given by $\theta = [x_1, y_1, x_2, y_2, x_L, y_L]^{\mathrm{T}}$, where L is the number of nodes to be localized. The observation vector X is formed by stacking the distance measurements $\hat{d}_{ij}$. Let M denote the size of X. We assume the distance measurement are Gaussian [62, 79], so the pdf of X is vector Gaussian. According to (6.1), we find that $CRLB = \{(1/\sigma^2)[G'(\theta)]^{\mathrm{T}}[G'(\theta)]\}^{-1}$, where σ^2 is the variance of each distance measurement error and $G'(\theta)$ is the $M \times 2L$ matrix whose mnth element is

$$G'(\theta)_{mn} = \begin{cases} \dfrac{x_i - x_j}{d_{ij}}, & \text{if} \theta_n = x_i; \\ \dfrac{x_j - x_i}{d_{ij}}, & \text{if} \theta_n = x_j; \\ \dfrac{y_i - y_j}{d_{ij}}, & \text{if} \theta_n = y_i; \\ \dfrac{y_j - y_i}{d_{ij}}, & \text{if} \theta_n = y_j; \\ 0, & \text{otherwise.} \end{cases} \tag{6.2}$$

The above result on CRLB is with the assumption that the location information of beacons is exact. When beacon nodes have location uncertainty, we can also characterize localization accuracy using a covariance bound that is similar to CRLB. Both these two bounds are tight in the sense that localization algorithms achieve these bounds for highly accurate measurements. In addition, according to (6.2), CRLB can be computed analytically and efficiently and avoid the need for expensive Monte Carlo simulations. The computational efficiency of CRLB facilitates to study localization performance of large-scale networks.

6.2.3 CRLB for One-Hop Localization

One-hop multilateration is the source of the location error that could be amplified by the iterative fashion of network localization. CRLB for multilateration exactly demonstrates how distance measurement errors and node geometry affect location accuracy.

Consider the one-hop localization problem: there are m reference nodes $v_1, v_2, \ldots, v_m$ and one node v_0 to be localized. From (6.1) and (6.2), we obtain

$$\sigma_0^2 = \sigma^2 m \left[\sum_{i=1}^{m-1} \sum_{j>i}^{m} \sin^2 \alpha_{ij} \right]^{-1}, \tag{6.3}$$

where σ_0^2 is the variance of the estimate location of v_0, α_{ij} is the angle between each pair of reference nodes (i, j). According to Eq. (6.3), the uncertainty of location estimate consists of two parts: the ranging error (σ_0^2) and the geometric relationship of references and the to-be-localized node (α_{ij}). Eliminating the impact of ranging errors, the error amplification effect caused by the node geometry has been demonstrated as the *geographic dilution of precision* (GDoP), which is defined as σ_0/σ.

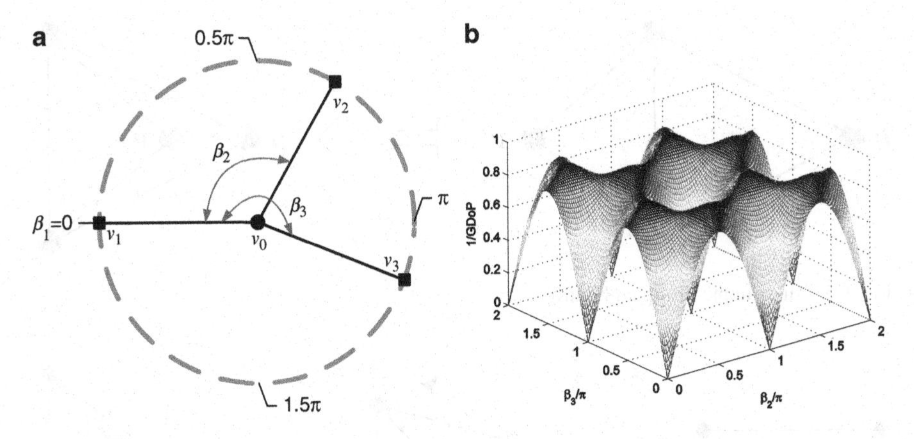

Fig. 6.2 The impact of node geometry on the accuracy multilateration

To gain more insights of GDoP, we consider a simplified case of multilateration, where the to-be-localized node v_O is put at the center of a circle and $m = 3$ reference nodes v_I, v_2, v_3 lie on the circumference of that circle, setting all references the same distance to v_O. Fixing v_I at $\beta_1 = 0$, according to the definition of GDoP, it becomes a function of the locations of v_2 and v_3, denoted by $\beta_2, \beta_3 \in [0, 2\pi]$, respectively. We plot the GDoP in Fig. 6.2 and conclude that different geometric forms of multilateration provide different levels of localization accuracy. In particular, in this circular trilateration, the highest location accuracy would be achieved if reference nodes are evenly separated, namely, $\beta_1 = 0, \beta_2 = \frac{2}{3}\pi$ and $\beta_3 = \frac{4}{3}\pi$.

6.3 Localization

AmbiguitiesIn the literature of graph realization problem, graph rigidity theory distinguishes between *flexible* and *rigid* graphs. Flexible graphs can be continuously deformed to produce an infinite number of different realizations preserving distance constraints, while rigid graphs have a finite number of discrete realizations. For rigid graphs, however, two types of discontinuous ambiguities exist, preventing a realization from being unique [62, 77]:

- **Flip.** Figure 6.3 shows an example of flip, where the two nodes in the middle create a mirror through which the position of v can be reflected without any change of inter-node distance.
- **Flex.** Discontinuous flex ambiguities occur when the removal of one edge allows the graph to be continuously deformed to a different realization and the removed edge can be reinserted with the same length. An example of flex ambiguity is illustrated in Fig. 6.4.

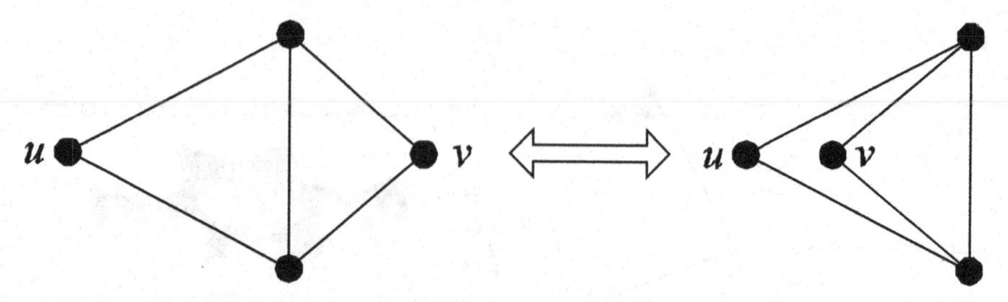

Fig. 6.3 An example of flip ambiguity

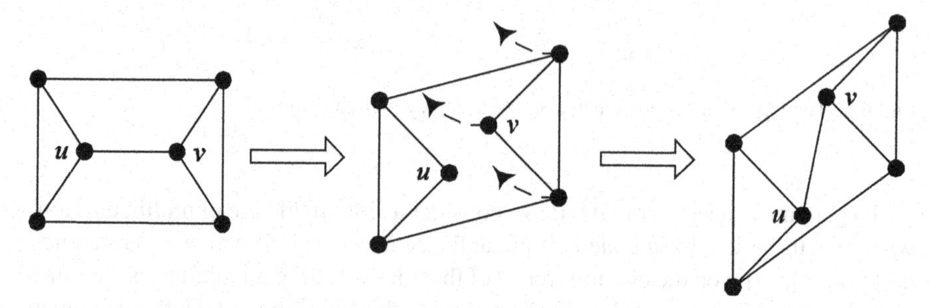

Fig. 6.4 An example of flex ambiguity

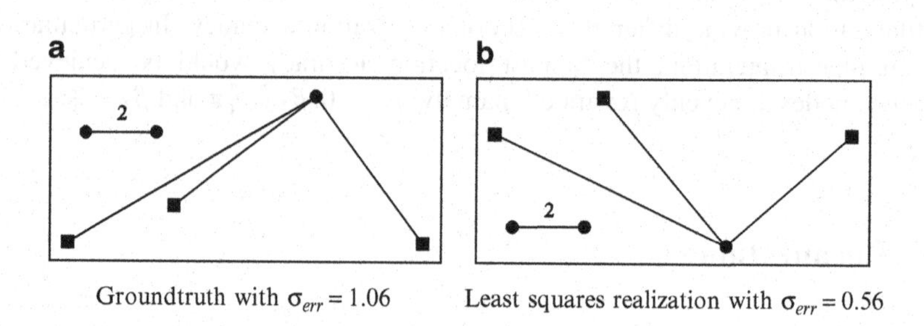

Fig. 6.5 An example of flip due to noisy distance measurements, where *black boxes* denote beacon nodes, and the *black circle* is the to-be-localized node

Graph rigidity theory [77] suggests ways of determining whether flip or flex ambiguities exist in a graph by checking global rigidity. However, this kind of criterion fails when distance measurements are noisy. Even when the underlying graph is globally rigid in the graph theoretic sense, realizations of the graph rarely satisfy the distance constraints exactly; furthermore, alterative realizations can exist and satisfy the constraints better than the correct one (the ground truth). An example of trilateration, which is a globally rigid structure, is illustrated in Fig. 6.5, where internode distance measurements are generated from a Gaussian distribution with a mean of the true distance and standard deviation $\sigma = 0.5$. Figure 6.5a is the ground truth realization with error metric $\sigma_{\text{err}} = 1.06$ which is defined as the average difference between the computed distances and the measured distances. Figure 6.5b is the least-squares realization which actually localizes the node at its mirror location with respect to the three beacon nodes, but with a much better error metric $\sigma_{\text{err}} = 0.56$.

Compared to flex ambiguities, flip ambiguities are more likely to occur in practical localization procedures and have attracted a lot of research efforts. In this section, we focus on the strategies of flip avoidance.

Moore et al. [62] outline certain criteria to select subgraphs to be used in localization against flip ambiguities due to noisy distance measurements. Rather than arbitrary quadrilaterals, they use "robust quadrilaterals" (robust quads) to localize nodes. As shown in Fig. 6.6, a robust quad consists of four subtriangles (ΔABC, ΔADC, ΔABD and ΔBCD) that satisfy

$$b \sin^2(\theta) > d_{\min} \tag{6.4}$$

where b is the length of the shortest side, θ is the smallest angle, and $d_{\min}$ is a predetermined constant according to the average measurement error. The idea is that the vertices of a quad can be placed correctly with respect to each other, i.e., without flip ambiguity. Moore et al. demonstrate that the probability of a robust quadrilateral experiencing internal flips given zero mean Gaussian measurement error can be bounded by setting $d_{\min}$ appropriately. In effect, $d_{\min}$ filters out quads that have too many positional ambiguities. The approximate level of filtering is based on the distance measurements. For instance, let $d_{\min} = 3\sigma$, then for Gaussian noise, we can bound the probability of flip for a given robust quadrilateral to be less than 1%, which poses minimal threat to the stability of the localization algorithm. Furthermore, these robustness conditions have a tendency to orphan nodes, either because they could not be localized by a robust quad or because their local map fail to overlap sufficiently with the global map. This tendency is acceptable because the orphaned nodes are likely to display large error. The drawback of this strategy is that under conditions of sparse networks or high measurement noisy, the algorithm may be unable to localize a useful number of nodes. Suggested in [27], there are other criteria that can better characterize the robustness of a given subnetwork against noisy ranging measurements.

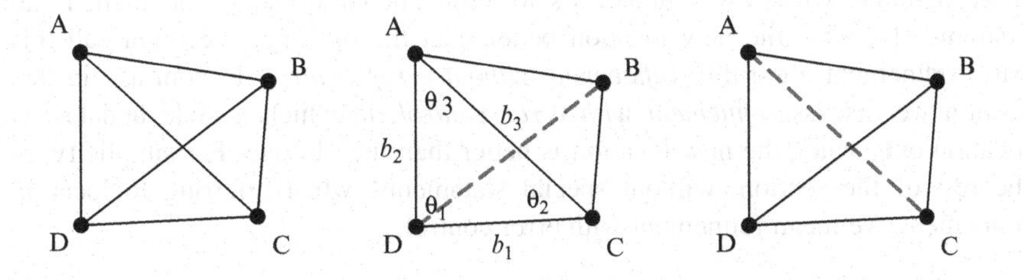

Fig. 6.6 Robust quadrilateral

Kannan et al. [80] propose another formal geometric analysis of flip ambiguity similar to robust quads [62]. Flip ambiguities are classified into two categories: *substantial flip ambiguity* and *negligible flip ambiguity*, based on the distance Δd between the two possible positions of the to-be-localized node. A flip ambiguity is substantial if $\Delta d \geq \delta_S$, some given bound. Otherwise, it is a negligible flip ambiguity. Instead of filtering out possible flip ambiguities as suggested in [62], they consider the identification of the substantial flip ambiguities only, because the location error introduced by negligible flip ambiguities is comparable to the uncertainty demonstrated by GDoP discussed previously. This strategy enables more nodes to be localized compared to robust quads. For a quadrilateral ABCD with known node positions A and B, Kannan et al. outline an algorithm to determine the region for the position of D such that node C can be uniquely localized using the measurements of the distances |AC|, |BC|, and |DC|.

6.4 Location Refinement

Since localization is often conducted in a distributed and iterative manner, error propagation is considered as a serious problem, in which nodes with inaccurate location estimates contaminate the localization process based on them. Existing studies [76, 79, 81] have demonstrated that location refinement is an effective technique to tackle this issue.

The basic location refinement requires nodes update their locations in a number of rounds. At the beginning of each round a node broadcasts its location estimate, receives the location information from its neighbors, and computes an LS-based multilateration solution to estimate its new location. In general, the constraints imposed by the distance to the neighbors will force the new location toward the ground truth location of the node. After a specified number of iterations or when the location update becomes small, the refinement stops and reports the final localization result.

The basic refinement algorithm is fully distributed, easy to implement, and efficient in communication and computation. An essential drawback of the basic refinement algorithm is that it is inherently unclear under what conditions the iteration would converge and how accurate the final solution would be, because in each round a node will update its location unconditionally, and there is no guarantee to make the new location better than the old one. We often call this basic refinement algorithm *refinement without error control*. In contrast, in this section we discuss *refinement with error control*, in which a node updates its location only when the new location is better than the old one. For simplicity, in the rest of the section, without special statements, when referring to location refinement, we mean refinement with error control.

6.4.1 A Framework of Location Refinement

To deal with error propagation, a number of location refinement algorithms have been proposed. In general, they are composed of three major components [79]:

1. ***Node registry***. Each node maintains a registry that contains the node location estimate and the corresponding estimate confidence (uncertainty).
2. ***Reference selection***. When redundant references are available, based on an algorithm-specified strategy, each node selects the reference combination achieving the highest estimate confidence (lowest uncertainty) to localize itself.
3. ***Registry update***. In each round, if higher estimate confidence (lower uncertainty) is achieved, a node updates its registry and broadcasts this information to its neighbors.

Algorithm 6.1 outlines the framework of location refinement, in which how to select appropriate reference combinations is the key step. Different strategies of addressing this issue lead to different location refinement algorithms.

6.4.2 Metrics for Location Refinement

Although GDoP characterizes the effects of node geometry on location estimate, it cannot be directly applied to the localization procedure due to the need of the ground truth location of each node. This is a challenging problem and has attracted a lot of research efforts.

Savarese et al. [81] propose a method that gives a confidence value to each node and weights one-hop multilateration results based on such confidence values. The estimate confidence is defined as follows. Beacons immediately start off with confidence 1; to-be-localized nodes begin with a low confidence (0.1) and raise their confidences at subsequent refinement iterations. In each round, a node chooses those reference nodes that will raise its confidence to localize itself, and sets its

Algorithm 6.1 A framework of location refinement

1: Each node holds the tuple (p, e), where p is the node location estimate, e is the corresponding estimate confidence (uncertainty).
2: Initialization step (optional):
   Each node computes an initialized location estimate.
3: In each round, nodes update their registries.
   do
      for all to-be-localized node t **do**
      examine local neighborhood $N(t)$
      select the best reference combination and compute the estimate location $\hat{p}_t$ and confidence $\hat{e}_t$
      decide whether to update the registry of t with the new tuple
   while the termination condition is not met.

confidence to the average of reference confidences after a successful multilateration. Nodes close to beacons will raise their confidence at the first iteration, raising in turn the confidences of node two hops away from beacons in the next iteration, etc. This strategy is based on the intuition that the estimated locations of nodes close to beacons are more reliable but puts little emphasis on node geometry.

Besides introducing the estimate confidence, Savarese et al. also consider the issue of ill-connected nodes, e.g., a cluster of n nodes with no beacons and connected to the main network by a single link, which are inherently hard or even impossible to locate. To detect non-ill-connected nodes, they adopt a heuristic criterion: a non-ill-connected node must have three edge disjoint paths to three distinct beacons. None of ill-connected nodes participate in the location refinement, which would make the algorithm convergence much faster.

By analyzing the effects of ranging errors and reference location errors on the estimated locations, Liu et al. [79] design a location refinement scheme with error management. Each node maintains information (p, e), where p is the estimated location, and e is the corresponding estimate error, a metric reflecting the level of uncertainty. At the beginning, each beacon is initialized with a registry (*beacon_loc*, 0), and the to-be-localized nodes are initialized as (*unknown_loc*, ∞). To handle errors, a robust LS (RLS) solution is adopted instead of the traditional LS solution $(A^TA)^{-1}A^Tb$ (discussed in Section 3.1) that gives

$$\hat{x}_t = \arg\min_x |Ax - b|^2$$

Let ΔA and Δb denote the perturbations of A and b, respectively. The RLS solution aims at

$$\hat{x}_t = \arg\min_x |(A + \Delta A)x - (b + \Delta b)|^2$$

With the assumption that ΔA and Δb are zero mean, the cost to minimize is

$$\begin{aligned}
\varepsilon &= E|(Ax - b) + (-\Delta b)|^2 \\
&= (x^TA^TAx - 2x^TA^Tb + b^Tb) + (x^TE[\Delta A^T\Delta A]x - 2x^TE[\Delta A^T] + E[\Delta b^T\Delta b]) \\
&= x^T(A^TA + E[\Delta A^T])x - 2x^T[A^Tb + E[\Delta A^T\Delta b]] + (b^Tb + E[\Delta b^T\Delta b])
\end{aligned}$$

Accordingly, the RLS solution is given by

$$\hat{x}_t = (A^TA + C_A)^{-1}[A^Tb + r_{Ab}]$$

where $C_A = E[\Delta A^T \cdot \Delta A]$ is the covariance matrix of perturbation of ΔA, corresponding to the uncertainties of reference locations, and $r_{Ab} = E[\Delta A^T \Delta b]$ is the correlation between ΔA and Δb. If ΔA and Δb are uncorrelated, the value of this term is 0. The analysis from [79] suggest that r_{Ab} is often negligible compared to the term A^Tb. Thus, the RLS solution becomes

$$\hat{x}_t = (A^T A + C_A)^{-1} A^T b. \tag{6.5}$$

Compared to the LS solution, the RLS solution uses the error statistics C_A as regularization, which would improve stability significantly when A is nearly singular or ill-conditioned. Based on the RLS solution, the location estimate error caused by noisy distance measurements can be expressed by

$$\begin{aligned} E|e_{\Delta b}|^2 &= E|(A^T A + C_A)^{-1} A^T \Delta b|^2 \\ &= E[\Delta b^T A (A^T A + C_A)^{-T} (A^T A + C_A)^{-1} A^T \Delta b] \\ &= \text{trace}[A (A^T A + C_A)^{-T} (A^T A + C_A)^{-1} A^T \text{Cov}(\Delta b)] \end{aligned}$$

Similarly, the error due to reference location uncertainty is

$$\begin{aligned} E|e_{\Delta a}|^2 &= E|(A^T A + C_A)^{-1} B \Delta a|^2 \\ &= \text{trace}[B^T (A^T A + C_A)^{-T} (A^T A + C_A)^{-1} B \text{Cov}(\Delta a)] \end{aligned}$$

where $a = (a_{11}, a_{21}, a_{n1}, a_{12}, a_{22}, a_{n2})^T$, a vector rearranging elements in matrix A, Δa is the perturbation of a because of location uncertainty, and B is a matrix satisfying $A^T b = Ba$, i.e.,

$$B \triangleq \begin{pmatrix} b_1 & b_2 & \dots & b_n & 0 & 0 & \dots & 0 \\ 0 & 0 & \dots & 0 & b_1 & b_2 & \dots & b_n \end{pmatrix},$$

in which b_1, b_2, b_n are elements in b. The total location error is the summation of these two terms, as they are assumed to be uncorrelated, i.e., $\hat{e} = E|e_{\Delta b}|^2 + \beta E|e_{\Delta a}|^2$, where β is a parameter to compensate for the over-estimation of the error due to a. A small value of β works well in practice [79].

By defining Quality of Trilateration (QoT) [76], the accuracy of trilateration can be characterized, enabling the comparison and selection among various geometric forms of trilateration. Assuming some probability distribution of ranging errors, probability tools are accordingly applied to quantify trilateration. The large value of QoT indicates the estimate location is, with high probability, close to the real location.

Let $t = \text{Tri}(s, \{s_i, i = 1,2,3\})$ denote a trilateration for a target node s based on three reference nodes s_i. The quality of trilateration t is defined as

$$Q(t) = \int_p \prod_{i=1}^{3} f_{s,s_i}(d(p, p(s_i)))dp, \ p \in \text{Disk}(p_t(s), R), \tag{6.6}$$

where $f_{s,s_i}(x)$ is the pdf of the distance measurement between s and s_i, and $p_t(s)$ is the estimated location based on trilateration t, and Disk(p, R) is a disk area centered at p with radius R. The parameter R is application specific for different accuracy requirements. To gain more insight of QoT, Yang et al. [76] provide some instances

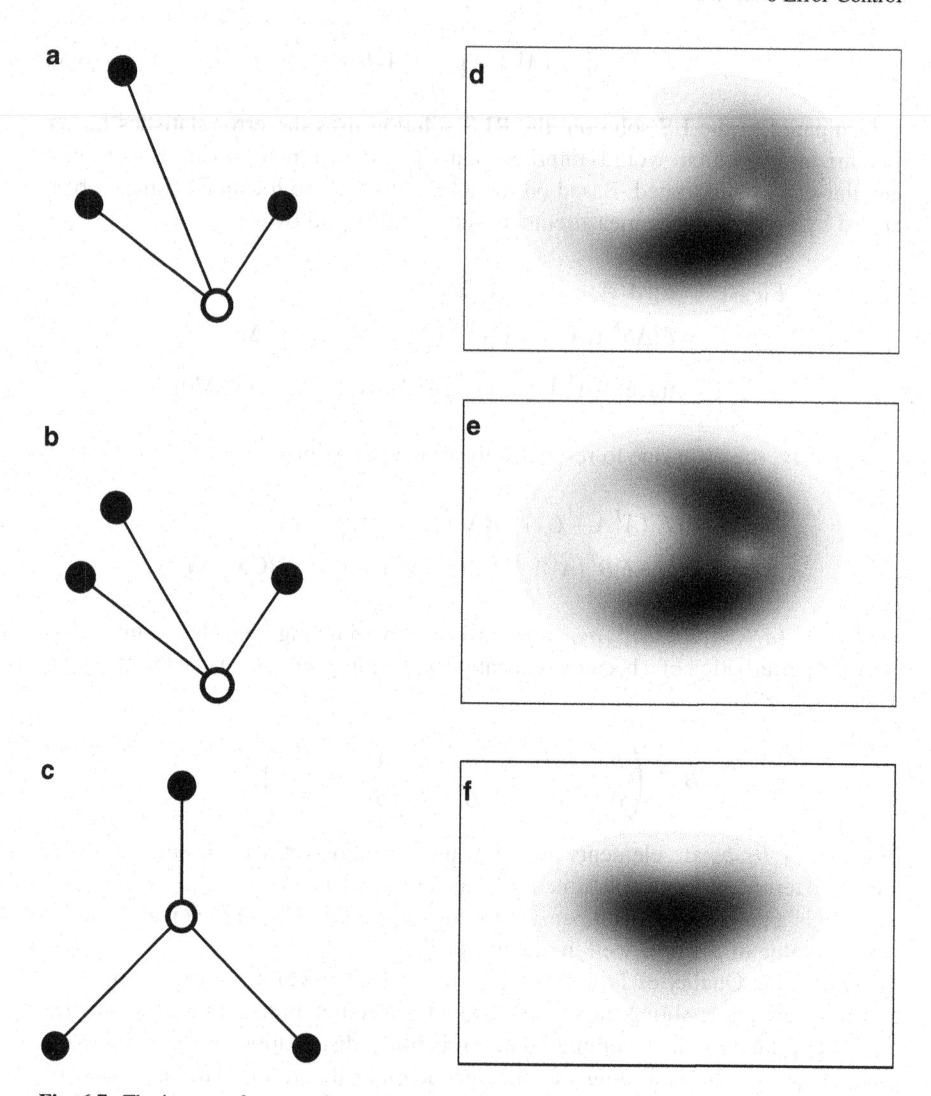

Fig. 6.7 The impact of geometric relations on QoT

to illustrate the impact of geometric relationship on QoT, shown in Fig. 6.7. Figure 6.7(a), (b), and (c) show the ground truths of three examples of trilateration. The black circles are the references and the white ones are the nodes to be localized. Based on the assumption that ranging measurements are with normal noises, the corresponding probability distributions are shown in Fig. 6.7. For the first instance, Fig. 6.7 displays the probability distribution of a general case. And for the second instance, Fig. 6.7(e) indicates a high probability of flip ambiguity as three references nodes are almost collinear. In the third instance, Fig. 6.7(f) plots a concentrated probability distribution which is accord with the fact that three references in Fig. 6.7(c) are well separated around the node to be localized.

Similar to [81], each node maintains a confidence associated with its location estimate. The confidence of s (based on t) is computed according to the confidences of references $C(s_i)$:

$$C_t(s) = Q(t) \prod_{i=1}^{3} C(s_i). \tag{6.7}$$

In each iteration, a to-be-localized node selects the trilateration that achieves the highest confidence to localize itself. Different from [81] that only takes the reference nodes reliability into account, QoT also considers the effects of geometry when computing confidence. Compared to conventional LS-based approaches, QoT provides additional information that indicates how accurate a particular trilateration is. Such difference enables QoT the ability of distinguishing and avoiding poor trilaterations that are of much location uncertainty.

6.5 Outlier-Resistant Localization

Compared with intrinsic errors, extrinsic errors are more unpredictable and caused by non-systematic factors. Especially in some cases, the errors can be extremely large due to the following factors:

- **Hardware malfunction or failure**. Distance measurements will be meaningless when encountering ranging hardware malfunction. Besides, incorrect hardware calibration and configuration also deteriorate ranging accuracy, which is not much emphasized by previous studies. For example, RSS suffers from transmitter, receiver, and antenna variability, and the inaccuracy of clock synchronization results in ranging errors for TDoA.
- **Environment factors**. RSS is sensitive to channel noise, interference, and reflection, all of which have significant impact on signal amplitude. The irregularity of signal attenuation remarkably increases, especially in complex indoor environments. In addition, for the propagation time based ranging measurements, e.g., TDoA, the signal propagation speed often exhibits variability as a function of temperature and humidity, so we cannot assume that the propagation speed is a constant across a large field.
- **Adversary attacks**. As location-based services are getting prevalent, the localization infrastructure is becoming the target of adversary attacks. By reporting fake location or ranging results, an attacker, e.g., a compromise (malicious) node, can completely distort the coordinate system. Different from the previous cases, the large errors here are intentionally generated by adversaries.

These severe errors can be seen as outliers of measurements. We classify the outlier-resistant approaches into two major categories: explicitly sifting and implicitly de-emphasizing. The explicitly sifting methods are usually based on the intuition that normal ranging measurements are compatible while an outlier is likely to be inconsistent with other normal and outlier rangings. By examining the inconsistency, we can identify and reject outlier measurements. In contrast, the implicitly de-emphasizing methods do not accept or reject a localization result by fixing a threshold, but employ robust statistics methods, for example, high breakdown point estimators and influence functions, to mitigate the negative effects of outliers.

6.5.1 Explicitly Sifting

The basic idea of outlier sifting is that the redundancy of geometric constraints can, to some extent, reveal the inconsistency of outlier ranging and normal ranging. Suppose m location references locating at p_i, $i = 1,2,\ldots,m$, are used to locate a target node by multilateration. Liu et al. [82] uses the mean square error ς^2 of the distance measurements as an indicator of inconsistency, i.e.,

$$\varsigma^2 = \frac{1}{m}\sum_{i=1}^{m}(\delta_i - |\bar{p}_0 - p_i|)^2, \tag{6.8}$$

where δ_i is the measured distance to the ith reference and $\bar{p}_0$ is the estimated location of the target node. A threshold-based approach is proposed to determine whether a set of location references is consistent. Formally, a set of location references $L = \{(p_i,\ \delta_i),\ i = 1,2,\ldots,m\}$ obtained at a sensor node is τ-consistent if the corresponding mean square error ς^2 satisfies $\varsigma^2 \leq \tau^2$.

Apparently, the threshold τ has significant impact on localization performance. The determination of τ depends on the measurement error model, which is assumed to be available. Based on the measurement error model, an appropriate τ is determined by performing simulation off-line. This threshold is stored at each senor node. In general, when the error model changes frequently and significantly, the fixed value of τ would degrade the performance. For simplicity, Liu et al. [82] assume the measurement error model will not change.

Given a set L of n location references and a threshold τ, it is desirable to compute the largest set of τ-consistent location references, because LS-based methods can deal with measurement errors better if there are more normal ranging results. The naive approach is to check all subsets of L with i location references about τ-consistency, where i starts from n and decreases until a subset of L is found to be τ-consistent or it is not possible to find such a set. Suppose the largest set of consistent location references consists of m elements. Then the sensor node has to perform LS-based localization at least $1 + \binom{n}{m+1} + \binom{n}{m+2} + \cdots + \binom{n}{n}$

times to figure out the right one. Although such an approach can provide the optimal result, it requires a large amount of computation when n and m are large numbers, which sometimes is unacceptable for resource constrained sensor nodes. To address this issue, Liu et al. [82] adopt a greedy algorithm, which is efficient but suboptimal. The greedy algorithm works iteratively. It starts with the set of all available location references. In each iteration, it checks whether the current set of location references is τ-consistent. If positive, the algorithm outputs the estimated location and stops. Otherwise, it considers all subsets of location references with one fewer location reference, and chooses the subset with the minimum mean square error as the input to the next iteration. Similar to the brute-force algorithm aforementioned, the greedy algorithm continues until it finds a set of τ-consistent location references or when it is not possible to find such a set. In general, through the greedy algorithm, the sensor node needs to perform LS-based localization for at most $1+n+(n-1)+\cdots+4$ times, which is much better than the brute-force algorithm.

Another way is to handle phantom nodes that claim fake locations. Hwang et al. [83] propose a speculative procedure, which can effectively and efficiently filter out phantom nodes. The filtering procedure is illustrated in Algorithm 6.2, where $Nbr(v)$ is the node set consisting of v and its neighbors, and E is used to keep consistent edges. G is initially empty. After computing the locations of all neighbors in the local coordinate system L by trilateration, for any two neighboring nodes j and k, if the difference between the measured distance and the computed distance is less than a threshold ε, the edge $e(j, k)$ is inserted into E. The threshold value ε depends on the noise in the ranging measurement. The largest connected cluster is regards as the largest consistent subset in the speculative plane L. This filter is done

Algorithm 6.2 Speculative filtering

for $i = 0$ to *iter* **do**
  node v randomly selects two neighbors u and w
  create local coordinate system L using v, u, w and their inter-distances $\hat{d}_{vu}$, $\hat{d}_{vw}$ and $\hat{d}_{uw}$
  initialize undirected graph $G(V, E)$
  create nodes v, u, w with locations p_v, p_u, p_w in V, respectively
  for each node $k \in Nbr(v)$ **do**
    calculate the location of k, p_k, in L by trilateration using p_v, p_u, p_w and $\hat{d}_{kv}$, $\hat{d}_{ku}$, $\hat{d}_{kw}$
    create node k with location p_k in V
  end for
  for each pair of nodes $j, k \in V$ and the distance $\hat{d}_{jk}$ **do**
    if $\left|\hat{d}_{jk} - \left|p_j - p_k\right|\right| < \varepsilon$ **then**
      create edge $e(j, k)$ in E
    end if
  end for
  find the largest connected cluster C and save it
end for
choose the one with the largest size among all saved C

iter times, where *iter* is determined by the application requirement, and the cluster with the largest size is chosen as the final result. The theoretical foundations of this strategy are the following two arguments:

- If the three pivots are honest nodes, the cluster output by Algorithm 6.2 contains no phantom nodes.
- If one pivot is a phantom node, the size of largest cluster is smaller than the one when none of pivots is a phantom node.

Departing from the two works previously discussed, which focus on security scenarios, in a recent work [84], Jian et al. propose a more general framework of sifting noisy and outlier distance measurements for localization. They formally define the problem of outlier detection for localization, and build the theoretical foundations based on graph embeddability and rigidity. Accordingly, an outlier detection algorithm is designed based on bilateration and generic cycles. Their results suggest the algorithm significantly improves the localization accuracy by wisely rejecting outliers. We discuss this work more detailed here.

Based on the grounded graphs associated with network instances, Jian et al. formulate normal ranging results and outlier ranging results as normal edges and outlier edges, respectively. In their error model of distance ranging, normal edges contain no ranging noise, while the measured distance of an outlier edge is an arbitrary continuous random variable. They argue that this assumption and abstraction is a good starting point to address the outlier detection problem. Through introducing an error threshold, their proposed algorithm can handle a more practical error model, where normal edges are with moderate ranging errors. Based on the normal edge and outlier edge model, the definition of outlier detection is straightforward: given a weighted graph $G = <V, E, W>$ consisting of normal and outlier edges, identify those outlier edges in G.

The theoretical foundations are built based on graph embeddability and rigidity. The first result provided by Jian et al. is:

Theorem 6.1. *Given a weighted grounded graph* $G = <V, E, W>$, *if* G *is unembeddable, the* E *contains at least one outlier edge.*

This result is intuitive: if G contains no outlier, then the ground truth is an embedding, and G cannot be unembeddable. Nevertheless, this is the best we can do for detecting outliers only based on ranging information. Formally, if G is embeddable, even G actually contains some outliers, we have no way to detect them.

Theorem 6.1 only provides a fine-granularity way to detect outliers, but cannot tell which are outliers and which are not. To address this issue, a concept of outlier disprovable is proposed:

Definition 6.1. *Given a weighted graph* $G = <V, E, W>$, G *is outlier disprovable if and only if the embeddability of* G *implies that it contains no outlier edge.*

Jian et al. prove the second result:

Algorithm 6.3 Outlier Detection Algorithm

```
for all redundantly rigid component H in G do
    if H is embeddable then
        mark every edge e ∈ H a normal edge
    else
        for all edge e ∈ H not marked a normal edge do
            mark e an outlier edge
        end for
    end if
end for
```

Theorem 6.2. *Given a weighted graph G, G is outlier disprovable if and only if G is redundantly rigid.*

A graph is rigid if it has no continuous deformation other than global rotation, translation and reflection while preserving distance constraints; otherwise, it is flexible. A graph is called redundantly rigid if it remains rigid after removing any single edge. Based on these two results, an outlier detection algorithm is designed as Algorithm 6.3. Different from [82] and [83], which are based on quadrilateral structures and require dense networks, Algorithm 6.3 pays more attention to exploring and utilizing the redundantly rigid topological structures, and thus, works properly in networks with moderate connectivity.

Algorithm 6.3 addresses the problem of outlier detection and identification theoretically. However, in practice, it suffers from the computational prohibitiveness, termed as combinational explosion, which is implied by the following result.

Theorem 6.3. *Let $G_1 = <V_1, E_1>$ and $G_2 = <V_2, E_2>$ be two redundantly rigid graphs with $|V_1 \cap V_2| \geq 2$. Then $G_1 \cup G_2$ is redundantly rigid.*

Suppose we have checked the embeddability of G_1 and G_2, both of which are redundantly rigid. According to Algorithm 6.3, we still need to check the embeddability of $G_1 \cup G_2$, which is actually implied by the checking results of G_1 and G_2 if $|V_1 \cap V_2| \geq 2$. To tackle this issue, Jian et al. introduce the concept of *generic cycle*, which is the minimally redundant rigidity.

Definition 6.2. *A graph $G = <V, E>$ with $|V| \geq 4$ is called a generic cycle if $|E| = 2|V| - 2$ and G satisfies*

$$i(X) \leq 2|X| - 3 \quad \text{for all } X \subset V \text{ with } 2 \leq |X| \leq |V| - 1$$

where $i(X)$ denotes the number of edges induced by X in G.

Based on generic cycles, another outlier detection algorithm is proposed, outlined by Algorithm 6.4, which avoids the computational prohibitiveness.

Algorithm 6.4 Edget-based Outlier Detection Algorithm.

```
for all e not marked in G do
    if there is a generic cycle H containing e is embeddable then
        mark every edge ∈ H, including e, a normal edge
    else
        mark e an outlier edge
    end if
end for
```

6.5.2 *Implicitly De-emphasizing*

(a) **What is Robust Statistics?**

In the literature of statistics, classical methods, e.g., mean and least squares, rely heavily on some idealized assumptions about input data sets, which are often not met in practice. Particularly, it is often assumed that the data residuals, i.e., the difference between the computed value and the input value, are normally distributed, or at least approximately. However, when there are outliers in the input data set, these methods often show very poor performance. Robust statistics is a theoretical framework concerning the outlier rejection problem, which provides alterative approaches to classical statistical methods in order to produce estimators that are not unduly affected by outliers. Two of the most common measures of robustness are *breakdown point* and *influence function*.

The breakdown point of an estimator is the fraction of data that can be given arbitrarily large values without giving an arbitrarily large result. For instance, it is obvious from the formula of the mean estimator, $\frac{1}{n}(x_1 + x_2 + +x_n)$, that if we hold x_1, x_2, x_{n-1} fixed and let x_n approach infinity, the statistic result also goes to infinity. In short, even one gross outlier can ruin the result of the mean estimator. Thus, such an estimator has a breakdown point of 0. In contrast, the median estimator can still give out a reasonable result when half of data goes to infinity. Accordingly, the breakdown point of the median estimator is 50%. The higher the breakdown point of an estimator, the more robust it is. Clearly, 50% is the theoretically highest breakdown value that can be achieved by an estimator, because if more than half of data is contaminated, it is impossible to distinguish the underlying distribution from the contaminating distribution.

Other than the breakdown point, the influence function is used to characterize the importance of individual data samples. A smaller absolute value of the influence function means the data item receives less weight in the estimation. The influence function is proportional to the derivative of the estimator. A robust estimator should have a bounded influence function, which does not go to infinity when the data value becomes arbitrarily large.

(b) **Robust Statistics Based Localization**

According to robust statistics [85], the least squares algorithm is sensitive to outliers, since its breakdown point is zero. One of the most commonly used robust fitting algorithms is the method of least median of squares (LMS), introduced by Rousseeuw et al. [85], which is adopted in [86] to design a robust localization algorithm. Instead of minimizing the summation of the residue squares, LMS minimizes the median of the residue squares, i.e., it estimates the location using

$$\bar{p}_0 = \arg\min_{p_0} med_i(\delta_i - ||p_0 - p_i||)^2, \tag{6.9}$$

where the parameters are defined in Section 6.5.1. In contrast to the least squares method, in which a single influential outlier may destroy the estimation, a single outlier has little effect on the objective function of LMS, and will not bias the estimate significantly. Results from [85] show that LMS has a breakdown point of 50%; in other words, LMS tolerates up to 50% outliers among all measurements and still outputs the correct estimate.

It is computationally prohibitive to get the exact solution of LMS. Rousseeuw et al. proposed an efficient and statistically robust alterative. First, using LS solution, we compute several candidate estimations according to random subsets of samples. The median of the residue squares for each candidate is then computed, and the one with the least median of residue squares is chosen as a tentative estimate. However, this tentative estimate is computed based on a small subset of data samples. As discussed previously, a better estimation can be achieved if more normal ranging results are included. To address this issue, the samples are weighted based on their residue for the tentative estimate, followed by a weighted least square fitting to get the final estimate. A simple threshold-based weighting strategy is as follows:

$$w_i = \begin{cases} 1, & \left|\frac{r_i}{s_0}\right| \leq \gamma; \\ 0, & \text{otherwise.} \end{cases} \tag{6.10}$$

where γ is a predetermined threshold, r_i is the residue of the i-th sample for the least median subset estimate $\bar{p}_0$, and s_o is the scale estimate given by [85] for the two dimensional estimated variable $\bar{p}_0$,

$$s_0 = 1.4826(1 + \tfrac{5}{n-2})\sqrt{\text{med}_i r_i^2(\bar{p}_0)}, \tag{6.11}$$

where n is the number of available samples. The term $(1 + \frac{5}{n-2})$ is used to compensate the tendency for a small scale estimate when there are few samples.

In summary, the LMS-based robust localization algorithm has the following steps:

- **Parameters selection**. Suppose n references are available. Choose an appropriate subset size m, the total numbers M of subsets, and a threshold γ.
- **Subsets generation**. Randomly draw M subsets of size m from the data set. Find the estimate $\bar{p}_0$ (using LS solution) for each subset. For each $\bar{p}_0$, calculate the median of residues r_i of each reference, $i = 1, 2, \ldots, n$.

- **Least median calculation**. Calculate $\bar{p}_0 = \arg\min_{p_0} med_i(\delta_i - |p_0 - p_i|)^2$ and residues with respect to $\bar{p}_0$, $r_i(\bar{p}_0)$.
- **Weights assignment**. Calculate s_0 based on Eq. (6.11) and assign weight w_i to each sample using Eq. (6.10).
- **Location estimation**. Do a weighted least squares fitting to all data to get the final location estimate.

The basic idea of LMS implementation is that, at least one subset among all randomly drawn subsets does not contain any outlier, and the estimate from this good subset will fit the inliers well. The chosen values of m and M have significant impact on the probability of such solution. The probability P of getting at least one good subset without outlier is calculated as follows. Assuming the contamination rate is ε, then

$$P = 1 - (1 - (1 - \varepsilon)^m)^M. \tag{6.12}$$

Given $m = 4$ and $M = 20$, the LMS algorithm is resistant up to 30% contamination with $P \geq 0.99$.

Inspired by robust statistics, the recent work SISR [87], from the perspective of influence function, analyzes the non-robustness of LS-based methods to outliers, and uses a residual shaping influence function to de-emphasize the "bad nodes" and "bad links" during the localization procedure.

The motivation of SISR is illustrated in Fig. 6.8, where different localization schemes would lead to different solutions. It is more desirable to get the uneven solution rather than the even solution, because E cannot be localized accurately in any case, given that it has large measurement error. Furthermore, since A, B, C and D could potentially be localized with great accuracy, a localization method that returns

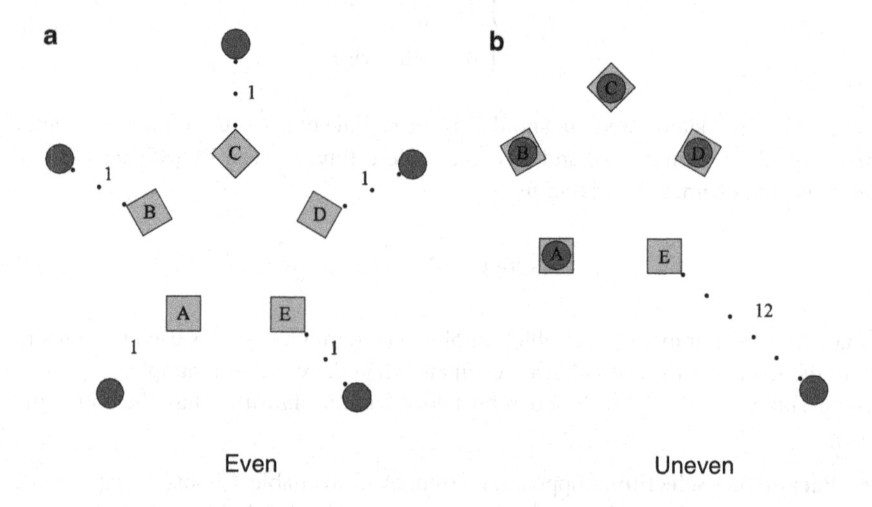

Fig. 6.8 Two possible solutions of nodes A, B, C, D and E, where E has large measurement errors. Squares indicate the ground truth locations; and circles the computed localization solutions. **(a)** The measurement errors from E is amortized over A, B, C and D. **(b)** Solutions for A, B, C and D are accurate, but that for E is very inaccurate

the uneven solution ought to de-emphasize the measurements of E to avoid contaminating the localization results of A, B, C and D. The conventional least squares method would find the even solution, as it does not distinguish between normal and outlier ranging measurements. To overcome this issue, SISR makes a key modification to the conventional least squares method: the residual function is shaped. As discussed previously, the influence function is proportional to the derivative of the estimator; in particular, here the estimator is the residual function. Thus, by shaping the residual function, the influence function is accordingly shaped in order to dampen the impact of outlier measurements and emphasize normal measurements. This modification make SISR find the uneven solution. The implementation of SISR is as follows.

Instead of optimizing the sum of squared residues, i.e., $F = \sum_{i,j} r^2(i,j)$, where $r(i,j)$ is the residue corresponding to edge (i, j), SISR solves the optimization problem of $F = \sum_{i,j} s(i,j)$, where

$$s(i,j) = \begin{cases} \alpha r(i,j)^2, & \text{if } |r(i,j)| < \tau \\ \ln(|r(i,j)| - u) - v, & \text{otherwise} \end{cases}, \tag{6.13}$$

where α, τ, u and v are parameters to be configured.

Figure 6.9 sketches the residual function of SISR, which has the following two properties:

- The shaping function increases with a smaller slope when the residual is large; in other words, the influence function of a measurement with large residual is smaller. In particular, the function dampens the impact of residuals larger than a threshold τ. This is called the wing-shaped section.
- The shaping function has a narrow and deep well for residuals close to 0. The normal measurements can therefore be emphasized by growing the shaped residuals more rapidly. This is called the U-shaped section.

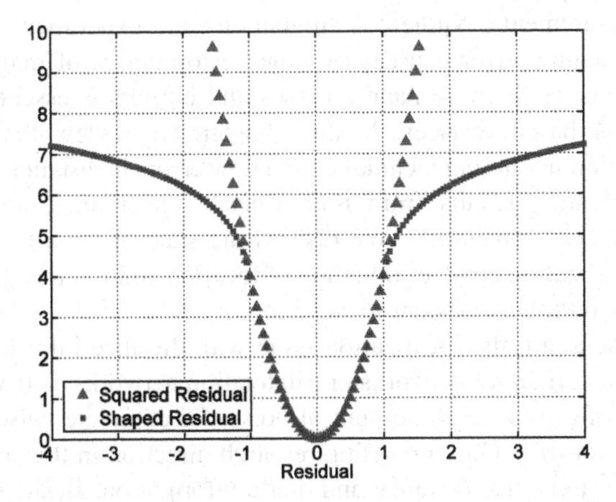

Fig. 6.9 Comparison between the standard squared residual used in conventional least squares and the shaped residuals used in SISR with $\alpha = 4$ and $\tau = 1$

The α and τ are two parameters that tune the shape of the SISR function in order to control its sensitivity to errors. The α is used to control the height of the U-shaped section, while τ controls its width. On the other hand, minimizing $F = \sum_{i,j} s(i,j)$ is a nonlinear optimization problem, and usually involves some iterative searching techniques, such as gradient descent or Newton method, to get the solution. Thus, it is necessary to make the SISR function piecewise-continuous and piecewise-differential at τ (apparently, it is piecewise-continuous and piecewise-differential at any other point). Accordingly, the other parameters u and v can be solved by

$$u = \tau - \frac{1}{2\alpha\tau}, \tag{6.14}$$

$$v = \ln\left(\frac{1}{2\alpha\tau}\right) - \alpha\tau^2. \tag{6.15}$$

Kung et al. [87] suggest that the value of τ has significant impact on the performance of the SISR estimator. On one hand, a small τ leads to more accurate localization results, while increases the probability of falling into an incorrect local minimum. On the other hand, a more permissive τ reduces this probability at the expense of localization accuracy. When τ approaches to infinity, SISR is actually reduced to the conventional least squares method. They design an iterative refinement scheme to exploit the above trade-off, and the proposed scheme works well in practice [87].

6.6 Summary

Although more ranging techniques are developed, noises and outliers are inevitable in distance measurements. Numerous simulations and experimental studies have suggested that ranging error can degrade the performances of many localization algorithms drastically. How to handle noises and outliers is essential for a wide range of location-based services. In this chapter, we review the measurement accuracies of different ranging techniques and how errors in distance measurements affect the localization results from four aspects: uncertainty, non-consistency, ambiguity and error propagation. We discuss the state-of-the-art works on characteristics of localization error, elimination of location ambiguities, location refinement schemes and outlier-resistant localization.

With no noise at all, the localization issue with distance information in dense networks, e.g., quadrilateral networks or trilateration networks, is trivial. However, in the presence of even a small amount of noise, for complete networks (graphs), localization is hard [88]. One promising research direction in this area is to adopt multimodal measurements, distance and angle information. Besides reducing the computational complexity of localization, multimodal measurements can provide more robustness to noises.

Chapter 7
Localization for Mobile Networks

7.1 Overview

In previous chapters, we have introduced many localization algorithms for wireless ad hoc and sensor networks. All those algorithms have a common assumption: nodes reside at fixed locations after being deployed, and localization can be done once for all. Recently, mobile networks are emerging because of the following two major reasons:

- *Passive motion.* In some applications, wireless nodes are attached to animals or people moving throughout an environment, e.g., in the ZebraNet [89], sensor nodes are placed on a sampled set of zebras. In addition, in some applications, e. g., ecosystem monitoring, wireless networks are deployed in dynamic environments, where nodes are in passive motion with the surroundings. For example, sensor nodes deployed on a sea surface are facing the motion of flows and waves [7, 90].
- *Active motion.* With the development of robotic platforms, wireless nodes can control underlying inexpensive robots and move autonomously for wide-area surveillance and reconnaissance. Mobility of nodes can improve network performance in a wide range of aspects, including increasing capacity [91], enhancing security [92], and improving connectivity [93].

Node mobility gives rise to new challenges on localization. The most straightforward and essential one is that localization is no longer a one-time task but a continuous and repeated procedure. In general, three ways exist for mobile network localization:

- Equip mobile nodes with global positioning system (GPS) receivers. However, as suggested when we discuss static networks, this solution faces nonavailable GPS signal and high hardware and energy costs, which make it not suitable for large-scale, low-cost mobile wireless networks. Nevertheless, it is more reasonable when adopted in mobile networks than static networks.
- Re-execute localization algorithms for static networks periodically to compute the real-time locations of mobile nodes. One important drawback of this strategy is that, in a highly dynamic environment where nodes move fast, localization

Y. Liu and Z. Yang, *Location, Localization, and Localizability: Location-awareness Technology for Wireless Networks*, DOI 10.1007/978-1-4419-7371-9_7,

algorithms need to be triggered frequently to meet accuracy requirements. In this case, significant energy and communication cost are inevitable.

- Design new techniques to deal with mobility. Localization algorithms for mobile networks often use some a priori information of node mobility, e.g., maximum velocity, to decrease energy and communication costs and improve location accuracy.

Compared to the first two trivial solutions, the third one is more appealing to large-scale mobile networks and recently has attracted a lot of research efforts. Thus, we focus on this kind of solution in this chapter. Without special statement, when referring to localization algorithms for mobile networks, we mean the third kind of solutions listed above in the rest of this chapter.

The remainder of this chapter is organized as follows. First, we introduce the Monte Carlo localization (MCL) algorithm, which casts the mobile localization problem as a Markov process, and employs sequential Monte Carlo (SMC) methods to resolve it. Then, we present the convex approximation localization (CAL) algorithms. Departing from MCL, CAL maintains a convex polygon or circle to approximate the potential location of each node rather than location samples. Both MCL and CAL focus on using beacon nodes to localize their neighboring nodes and propagate global location information, and update new location based on sequential measurements. We also discuss the moving-baseline localization (MBL) algorithm, whose emphasis is to construct a globally consistent view of the network from the perspective of each individual node, i.e., distances and velocities of other nodes with respect to some node. Finally, some techniques for universal localization (locating both static and mobile nodes simultaneously) are depicted.

7.2 Monte Carlo Localization

7.2.1 Particle Filtering

In applications where the state of a system has to be estimated from some observations, the system can be formulated by using a Bayesian model in which the posterior distribution of the state of the system is only determined by the current observations and state of the system. In dynamic systems, observations arrive sequentially, and it is therefore required to update the posterior distribution of the system state upon arrival of new observations. Formally, the state of the system $\{x_t\}_{t=0,1,2,\ldots}$ is modeled as a Markov process with initial distribution $p(x_0)$ and transition prior $p(x_t \mid x_{t-1})$. The observations $\{y_t\}_{t=1,2,\ldots}$ are assumed to be conditionally independent given $\{x_t\}$ and with marginal distribution $p(y_t \mid x_t)$.

If the initial state and the observations can be modeled by a linear Gaussian state-space model, it is possible to derive an exact analytical expression to compute the evolving sequence of posterior distributions. This is the well-known and widely accepted Kalman filter. However, the states and observations of real systems are often very complex, typically involving elements of non-Gaussianity, high dimensionality, and nonlinearity, which usually preclude analytic solution. To address this challenge, particle filters, also known as sequential Monte Carlo (SMC) methods, are widely employed in practice, which are a set of simulation-based techniques providing a convenient approach to compute the posterior distributions in non-Gaussian environments.

The key idea of particle filtering is to represent the state distribution of a system by a set of N weighted samples:

$$p(x_t|y_{1,2,\ldots,t}) \approx \{x_t^i, w_t^i\}_{i=1,2,\ldots,N}, \tag{7.1}$$

where $p(x_t|y_{1,2,\ldots,t})$ is the posterior distribution of the system state at time t given observations $\{y_k\}_{k=1,2,\ldots,t}$, x_t^i is a sample of x_t, and w_t^i is the normalized importance weight associated with x_t^i. The number of samples maintained is an essential system parameter: a minimum of samples should be available so that the set of samples converges to the posterior distribution. A number of particle filters have been proposed in the literature. Here, we illustrate the major steps of particle filtering by introducing the most typical one, bootstrap filter. Actually, all SMC-based algorithms to be discussed in this chapter employ bootstrap filter since it is easy to implement and computationally efficient.

Algorithm 7.1 depicts the three steps of bootstrap filter: initialization, importance sampling, and selection. In the initialization step, N samples are randomly selected according to the initial distribution of the system, $p(x_0)$. In the second step, based on the transition prior $p(x_t \mid x_{t-1})$ and N samples x_{t-1}^i representing the previous

---

Algorithm 7.1 Bootstrap filter

1 Initialization step: $t = 0$
  for $i = 1$ to N **do**
    sample $x_0^i \sim p(x_0)$
  end for
  set $t = 1$
2 Importance sampling step:
  $W = 0$
  for $i = 1$ to N **do**
    sample $\tilde{x}_t^i \sim p(x_t|x_{t-1}^i)$ and set $\tilde{x}_{0:t}^i = (\tilde{x}_{0:t-1}^i, \tilde{x}_t^i)$
    calculate the importance weight $\tilde{w}_t^i = p(y_t|\tilde{x}_t^i)$
    $W = W + \tilde{w}_t^i$
  end for
  for $i = 1$ to N **do**
    $\tilde{w}_t^i = \tilde{w}_t^i / W$
  end for
3 Selection step:
  resample with replacement N particles $\{x_{0:t}^i\}_{t=1,2,\ldots N}$ from the set $\{\tilde{x}_{0:t}^i\}_{t=1,2,\ldots N}$ according to the importance weights
  set $t \leftarrow t + 1$ and go to Step 2

---

state x_{t-1}, N new samples x_t^i are chosen to represent the current state x_t. The importance weights w_t^i associated with x_t^i are computed based on the marginal distribution $p(y_t|x_t)$ and normalized. In the last step, after resampling according to the importance weights, the particles with small weights are eliminated while those with high weights are multiplied. The finalized N samples are with the same weight, i.e., $1/N$.

7.2.2 Sequential Monte Carlo Localization

From Algorithm 7.1, three distribution functions that are essential to bootstrap filter exist: the initial distribution $p(x_0)$, the transition prior $p(x_t|x_{t-1})$, and the marginal distribution $p(y_t|x_t)$. We discuss these three functions in the context of network localization. Although the Monte Carlo methods can apply to range-based localization for mobile networks [94], we focus on range-free localization here for the simplicity of discussion. The principles presented can be directly applied to the range-based localization.

Initial distribution. In general, nodes initially have no information about their locations except beacons. Therefore, the distribution $p(x_0)$ can be first modeled as a uniform distribution over the whole field of interest. One way to improve the performance of MCL, especially the convergence speed, is to use expensive localization algorithms for static networks, such as manual configuration, to provide relatively accurate $p(x_0)$.

Transition distribution. In the MCL framework, the motion of nodes is modeled as a Markov process: the location of a node at time t, x_t, is only determined by x_{t-1}, its location at time $t-1$. Generally, other than knowing its upper bound speed v_{max} (probably as a system configuration parameter), a node is unaware of its moving speed and direction. In other words, for each node, x_t must be contained in the circular region centered at x_{t-1} with radius v_{max}. It is also assumed that node speed is uniformly distributed in the interval $[0, v_{max}]$. Accordingly, the transition distribution $p(x_t|x_{t-1})$ is given by

$$p(x_t|x_{t-1}) = \begin{cases} \frac{1}{\pi v_{max}^2}, & \text{if } d(x_t, x_{t-1}) \leq v_{max}, \\ 0, & \text{otherwise,} \end{cases} \tag{7.2}$$

where $d(x_t, x_{t-1})$ denotes the Euclidean distance between x_t and x_{t-1}. This distribution function reflects the fact that the unknown motion of a node increases the uncertainty of its location. The larger the value of v_{max} is, the more uncertainty is introduced in each step. If an accurate mobility model is available, such as an accurate moving velocity or orientation, the transition distribution can be adjusted accordingly to provide better predictions.

Marginal distribution. The marginal distribution $p(y_t|x_t)$ represents the relationship between the observation y_t and the system state x_t at time t. Since we focus on range-free localization here, the only information available to each node is the existence of its neighbors. The 1-hop neighbors of a node u are nodes that can communicate with u directly. The 2-hop neighbors are nodes that communicate with at least one of 1-hop neighbors of u directly but cannot communicate with u directly. We can extend these definitions to k-hop neighbors. In the unit disk graph (UDG) model with communication radius r, for a k-hop neighbor v of u, the following geometric constraints is satisfied:

$$\begin{aligned} 0 \leq |\pi(v) - \pi(u)| &\leq r \text{ if } \ k = 1, \\ r < |\pi(v) - \pi(u)| &\leq k{*}r \text{ if } \ k \geq 2, \end{aligned} \tag{7.3}$$

where π (u) and π (v) denote the physical locations of u and v, respectively. Apparently, the greater the value of k is, the more communication cost and larger localization latency would be introduced, and the geometric constraints become less useful. Therefore, in practice, only 1-hop and 2-hop neighbors are used for location estimation.

Different schemes on using 1- and 2-hop neighbors lead to different marginal distributions. In particular, communication and computation cost trades off with the localization accuracy. Here, we review two existing works on MCL. The work presented in [95] relies on only 1- and 2-hop beacons. On the one hand, it is efficient in terms of communication and computation; on the other hand, abundant beacons are required to provide accurate location results. In contrast, the MSL in [96] uses information from all 1- and 2-hop neighbors, i.e., including beacon nodes and ordinary nodes.

Let $\{x_t^i\}_{i=1,2,\ldots,N}$ be the sample set representing the location of a node u at time t obtained based on $p(x_t|x_{t-1})$. Let S and T denote the sets of beacons which are 1-hop neighbors and 2-hop neighbors of u, respectively. In [95], the marginal distribution of u is given by

$$p(y_t|x_t^i) = \begin{cases} \frac{1}{N}, & \text{if } \forall s \in S, 0 \leq |\pi(s) - x_t^i| \leq r, \text{and}, \forall s \in T, r < |\pi(s) - x_t^i| \leq 2r, \\ 0, & \text{otherwise}, \end{cases} \tag{7.4}$$

where N is the number of samples maintained by the system as discussed previously. Intuitively, not all samples in $\{x_t^i\}_{i=1,2,\ldots,N}$ would have nonzero weights. To address this issue, particle filtering adopted in [95] makes some revisions at the "importance sampling" step of Algorithm 7.1: instead of generating N samples, it repeats the sampling procedure until there are N samples with nonzero weights. Since all the N surviving samples are equally weighted, the selection step of Algorithm 7.1 is actually eliminated.

Since information from all 1- and 2-hop neighbors is used, it is very complicated to compute the marginal distribution [96]. Let ST denote all 1-hop and 2-hop neighbors of node u. For each sample $x_t^i, i = 1, 2, \ldots, N$ chosen for u, $p(y_t|x_t^i)$ is computed by

$$p(y_t|x_t^i) = \prod_{s \in ST} w'_{x_t^i}(s), \tag{7.5}$$

where $w'_{x_t^i}(s)$ is the partial weight associated with s, one 1-hop or 2-hop neighbor of node u. If s is a beacon, $w'_{x_t^i}(s)$ is calculated using (7.4), except substituting 1 for $1/N$. Otherwise, let $\{s_i\}_{i=1,2,\ldots,N}$ denote the sample set for s, and $w'_{x_t^i}(s)$ is computed as follows: if s is a 1-hop neighbor, then $w'_{x_t^i}(s) = \sum_{s_j:|s_j - x_t^i| \leq r + v_{\max}} p(y_t|s_j)$; if s is a 2-hop neighbor, then $w'_{x_t^i}(s) = \sum_{s_j:r - v_{\max} \leq |s_j - x_t^i| \leq 2r + v_{\max}} p(y_t|s_j)$. MSL in [96] employs a threshold to eliminate those samples with low importance weights. The threshold adopted in MSL is $\beta = (0.1)^k$, where k is the number of 1-hop and 2-hop neighbors of a node.

7.3 Convex Approximation Localization

Now, we discuss another kind of localization for mobile wireless networks: convex approximation localization (CAL). Departing from MCL, in which a set of discrete points is maintained to represent the location candidate region of each node, CAL employs a convex polygon or circle to approximate the location region that contains the physical location of a node. The approximation simplifies the locate region update procedure due to node motions and decreases network communication cost. This scheme, however, cannot make use of nonconvex constraints [97]. One typical example of nonconvex constraint is the negative information [98], e.g., a node does not hear from a beacon located at somewhere. In general, however, the negative information plays an important role in refining location estimations. In the rest of the section, we discuss two existing convex approximation localization algorithms: one is convex polygon based, and the other one is circle based.

In [97], each node maintains a convex polygon that represents its location candidate region. At any time, the polygon associate with a node is sufficiently large to contain the physical location of the node, which is the key invariant of this algorithm. The centroid of the polygon is used as the estimated location of the node, and the size of the polygon can serve as a measure of the estimated uncertainty. Since it is often assumed that the radio range of a node is a perfect circle, how to represent a circle is critical for polygon-based localization. Figure 7.1a illustrates an example of using regular polygon to approximate a perfect circle. The major benefit of using polygons is that the intersection of convex polygons can be efficiently computed, compared to circles and curves.

The CAL algorithm proposed in [97] consists of the following three steps:

1. *Initialization.* Beacons start with small regular polygons to approximate a circle centered at the location of the beacon and with radius ε, whose value is chosen to

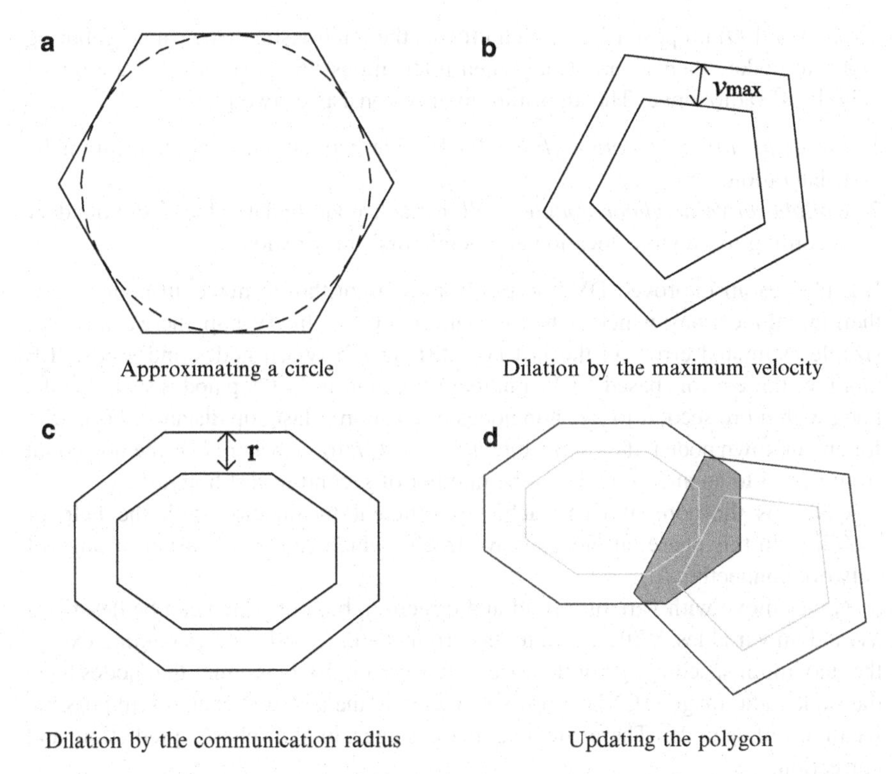

Fig. 7.1 Convex polygon approximation. (**a**) Approximating a circle, (**b**) dilation by the maximum velocity, (**c**) dilation by the communication radius, (**d**) updating the polygon

reflect the uncertainty associated with that estimated location. Since the to-be-localized nodes have no nontrivial a priori knowledge about their locations, they use the whole field of interest as their initial region.

2. *Dilation*. There are two kinds of dilation at each time step. The first one is due to the mobility of nodes. Each to-be-localized node dilates its polygon outward by v_{max} (illustrated in Fig. 7.1b), the maximum velocity. This dilation maintains the invariant that the true location of the node is in the polygon. The second one is to dilate the polygon by r (illustrated in Fig. 7.1c), the radio range. After two dilations, a node informs the final polygon (the candidate region) to its neighbors.
3. *Update*. Each node receives a set of polygons from its neighbors and then computes the intersection of those received polygons and its own polygon. The result is the new location polygon for that node. Figure 7.1d shows such a procedure. Intersecting a convex polygon with other convex polygons yields a smaller convex polygon with high probability.

CAL has the following salient features: free range, fully distributed, and no need for routing infrastructure, computationally efficient, etc.

Xi et al. [99] propose EUL, which utilizes the relationship between neighboring nodes to update their locations and then filter impossible positions that are out of neighbors' radio range. The algorithm mainly consists of two phases:

1. *Initial location estimation (ILE)*. Each node estimates its initial location by trilateration.
2. *Collaborative neighbor update (CNU)*. Each node updates location boundary according to previous location and neighbors' information.

ILE utilizes an improved DV-hop positioning algorithm to make all nodes know their initial locations, which is the foundation of CNU. In DV-hop, there always are sizable estimated errors in the last hop distances between nodes and seeds. ILE modifies these errors based on the number of second-to-last hop nodes such that the node with more second-to-last hop nodes has a shorter last hop distance. Formally, for an unknown node i, they substitute $h_i - 1 + 1/t_i$ for h_i, where h_i is the hop count from node i to an anchor and t_i is the number of second-to-last hop nodes.

CNU, as the core of EUL, achieves efficient localization with the help of mobility. In this phase, all nodes move freely, which causes a constant change of network connectivity.

Nodes move with variable speed and direction, both of which can be described as random variables. With regard to its current status, a node knows nothing except the maximum speed $v_{\max}$. For the ease of discussion, EUL assumes that nodes have the same radio range r. CNU aims to characterize the area that contains all possible locations of a node. There are two main stages in this phase: prediction and correction.

In the prediction stage, nodes estimate their current locations according to their previous locations and $v_{\max}$. At first, a node predicts possible location set S_1 in a circle whose center is its initial location estimated in ILE and radius is $v_{\max}$. In a subsequent time slot t, S_t is concentric with S_{t-1} are concentric circles, and radius of $v_{\max}$ greater than S_{t-1}. Figure 7.2a shows the relation between S_t and S_{t-1}.

In the correction stage, collaborating with neighbors, each node filters impossible locations. If a node has neighbors, it should be in the overlap of these neighbors' radio range including both new neighbors and constant neighbors. Thus the possible

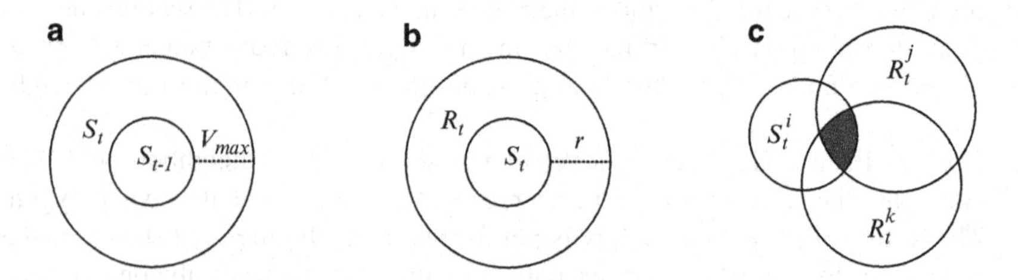

Fig. 7.2 Collaborative neighbors update. (**a**) The prediction of a node's location range; (**b**) the prediction of a node's radio range; (**c**) the correction of a node's location range

radio range should be estimated before filtering. Figure 7.2b shows the relationship between a node's possible radio range R_t and S_t. The node estimates R_t according to S_t and r similar to location range prediction process. Each node computes the overlap region between its possible location set and neighbors' possible radio range as its new location set.

There is a knotty problem that the boundary of a node's location set is irregular. It is hard to describe and compute when the number of neighbors is large. Approximation by maximum inscribed circle (MIC) is an effective method to regularize boundary. The center of MIC is the intersection point of circles, which have common centers with S_t or R_t. That is,

$$\begin{cases} (x_1 - a_1)^2 + (x_2 - b_1)^2 - (x_3 - c_1)^2 = 0 \\ (x_1 - a_2)^2 + (x_2 - b_2)^2 - (x_3 - c_2)^2 = 0 \\ \vdots \\ (x_1 - a_n)^2 + (x_2 - b_n)^2 - (x_3 - c_n)^2 = 0, \end{cases} \tag{7.6}$$

where (a, b) and c denote the center and radius of S_t or R_t, and (x_1, x_2) and x_3 are the center and radius of MIC, respectively. When n is greater than 3, the equation set is a nonlinear overdetermined set of equations, which requires nonlinear optimization methods to approximate. Nonlinear least-squares method is used to fit m observations with a model that is nonlinear in n unknown quantities ($m > n$). EUL [99] uses the Gauss–Newton method which is based on linear approximation of the objective function to approximate an optimum solution, where the Jacobian, J, is a function of constants:

$$J(x_k)^{\mathrm{T}} J(x_k) + \nabla f(x_k) = 0. \tag{7.7}$$

7.4 Moving-Baseline Localization

Moving-baseline localization (MBL) [100] deals with the absence of a fixed reference frame. The MBL problem arises when a group of nodes moves in an environment where no external coordinate reference is available. The goal of MBL is to enable each node to infer the spatial relationship and motion of all other nodes with respect to itself.

To model the pairwise interactions among nodes in a mobile network, a concept of dynamic network is introduced in [100], in which an edge between nodes i and j exists if and only if they can exchange information, i.e., it is assumed that when edge ij exists, a discrete sequence of range measurements $r_{ij}(t)$ is available at node i,

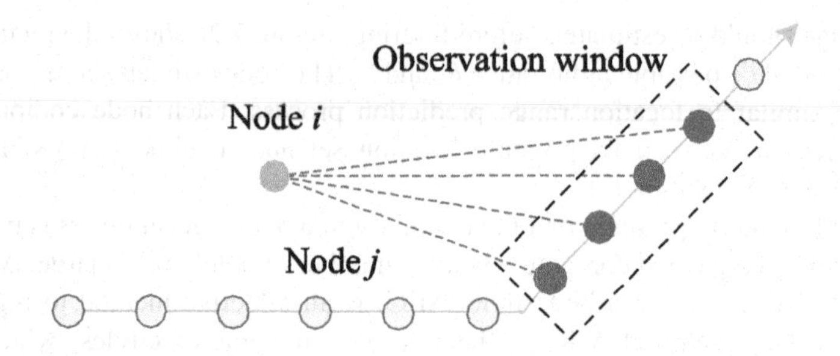

Fig. 7.3 Time-series range data $r_{ij}(t)$

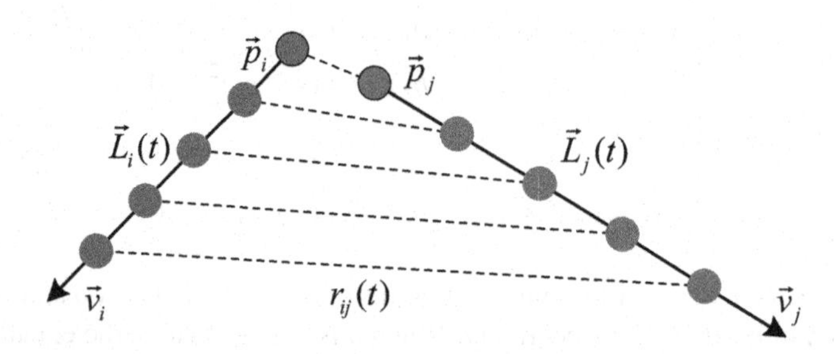

Fig. 7.4 MBL recovers four degrees of freedom per node

describing the measured range from node i to node j at time t observed by node i, as illustrated in Fig. 7.3.

Starting from the simplest case of MBL, where each node is moving along a straight-line path at a constant speed, recovering node trajectories can be cast as a low-dimensional optimization, as shown in Fig. 7.4. Formally, four degrees of freedom per node need to be recovered: the values of $\vec{p}_i$ and $\vec{v}_i$ in the following expression

$$\vec{L}_i(t) = \vec{p}_i + t \cdot \vec{v}_i, \tag{7.8}$$

where $\vec{p}_i$ and $\vec{v}_i$ represent the origin (corresponding to two degrees of freedom) and the velocity vector (corresponding to the other two degrees of freedom) of the motion of node i, and $\vec{L}_i(t)$ is the location of that node at time t. Accordingly, the distance ranges $r_{ij}(t)$ between nodes i and j lie on a hyperbola is defined by

$$r_{ji}^2(t) = m_{ji}^2 + (t - t_{ji}^c)^2 s_{ji}^2, \tag{7.9}$$

where t_{ji}^c denotes the time when nodes i and j are closest to each other, and m_{ji} represents the node separation distance at t_{ji}^c, and s_{ji} is the relative speed, i.e., $\|\vec{v}_j - \vec{v}_i\|$. These parameters are illustrated in Fig. 7.5. Clearly, as long as $H_{ji} = (s_{ji}, t_{ji}^c, m_{ji})$ is available, the distance r_{ji} between nodes i and j at any specific time can be computed by (7.9), as shown in Fig. 7.5. Because not any pair of nodes i

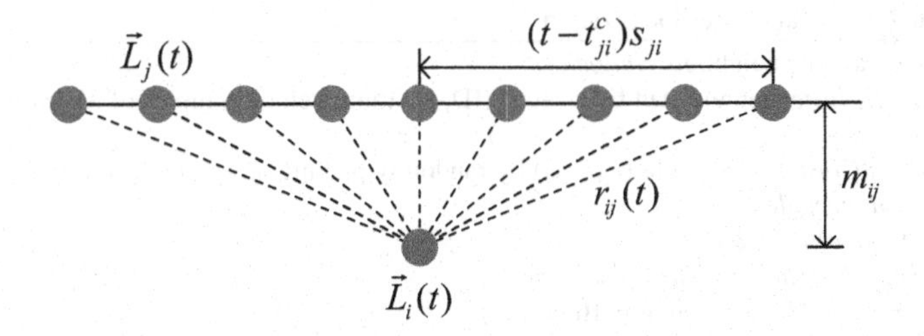

Fig. 7.5 Distance ranges lie on a hyperbola

and j is connected in the dynamic network, to achieve the goal of enabling each node to infer the spatial relationship and motion of all other nodes with respect to itself, an optimal global motion solution is needed to be reconstructed from all available H_{ji}, which involves the following steps.

Hyperbola estimation. The basic block of MBL algorithm is to estimate the motion hyperbola parameters $H_{ij} = (s_{ij}, t^c_{ij}, m_{ij})$ from time-stamped range measurements. Based on the motion model, i.e., (7.9), given a sequence of n discrete distance range measurements between two nodes, $(r_z, t_z), z = 1, 2, \ldots, n$, the following quadratic model is considered:

$$r_z^2 = \gamma t_z^2 + \beta t_z + \alpha + \varepsilon_z. \tag{7.10}$$

Instead of using parametric regression methods like the ordinary least-squares estimator, which are sensitive to data containing significant noise and outliers (modeled as ε_z), to estimate $\hat{\gamma}$, $\hat{\beta}$, and $\hat{\alpha}$, nonparametric robust quadratic fitting is adopted [100]. This method performs well even when the error distribution associated with the data is not normal. When $n \geq 3$, $\hat{\gamma}$, $\hat{\beta}$, and $\hat{\alpha}$ can be determined, and then the motion hyperbola parameters $H = (s, t^c, m)$ are calculated by

$$\hat{s} = \sqrt{\hat{\gamma}},\ \hat{t}^c = \hat{\beta}\big/(-2\hat{\gamma}) \text{ and } \hat{m} = \sqrt{\hat{\alpha} - \hat{\beta}^2\big/(4\hat{\gamma})}. \tag{7.11}$$

Based on the recovered parameters $H = (s, t^c, m)$, the hyperbola illustrated in Fig. 7.5 can be reconstructed. In general, more samples can improve the accuracy of the estimation. After computing the motion hyperbola parameters, each node shares this information with its neighbors to estimate local clusters.

Path estimation geometry. The estimated motion hyperbola parameters only capture the relative position and motion of a pair of nodes. To build a local cluster, three relations among nodes i, j, and k are needed to infer the relative motion of the node triangle, just like using distance information of three nodes to build a local coordinate system. From the perspective of node i, this problem can be cast as fixing i at the origin and determining the motions of j and k in the frame of i. Analysis in [100] shows that three motion relations are enough to tackle this problem.

Algorithm 7.2 Local cluster localization

Input: Node i and its neighbors *Neighbors*
Output: *LocalCluster* represented by a set of (ID, position, velocity) tuples of *Neighbors*
$doneNodes = \phi$
Initialize *LocalCluster* as triangle (i, j_0, k_0) by randomly picking j_0, k_0 from *Neighbors*
Add j_0, k_0 to *doneNodes*
for node $j \in doneNodes$ **do**
  for node $k \in Neighbors - doneNodes$ **do**
    if H_{ji}, H_{jk} and H_{ik} are available **then**
      Construct triangle (i, j, k)
      Merge (i, j, k) into *LocalCluster*
      Add k to *doneNodes*
    end if
  end for
end for

Local Cluster Localization. After constructing each triangle in its own frame, each node calculates the motion of its neighbors following a procedure analogous to chained trilateration. The algorithm for local cluster localization is outlined in Algorithm 7.2.

Global View Construction. To get a global view of a dynamic network, we can repeatedly find the best alignment for each pair of local clusters, i.e., merging two local clusters consistently, that share three or more noncollinear nodes, each time adding a local cluster to the major component. This problem is formulated as the absolute orientation problem, which can be solved efficiently by the eigendecomposition. Consider the problem of aligning cluster 2 to cluster 1. Let S denote the common nodes between cluster 1 and cluster 2. For $i \in S$, we use $(P_i^{(1)}, V_i^{(1)})$ and $(P_i^{(2)}, V_i^{(2)})$ to represent the position and velocity of node i in cluster 1 and cluster 2, respectively. The Euclidean transformation from $P^{(2)}$ to $P^{(1)}$ is recovered by minimizing the sum of squared residuals:

$$(\bar{R}, \bar{T}) = \arg\min_{R,T} \sum_{i \in S} \left\| P_i^{(1)} - R(P_i^{(2)}) - T \right\|^2, \tag{7.12}$$

where R corresponds to a rotation, while T is a translation. After solving for $(\bar{R}, \bar{T})$ by the eigendecomposition method, we solve

$$\bar{V} = \arg\min_{V} \sum_{i \in S} \left\| V_i^{(1)} - \bar{R}(V_i^{(2)}) - V \right\|^2, \tag{7.13}$$

to get the velocity offset $\bar{V}$ that best shifts velocities in cluster 2 to align with those of cluster 1.

7.4.1 Techniques for Universal Localization

Universal localization refers to the localization algorithms that can be applied to both mobile and static wireless networks. For static networks or slightly mobile networks, i.e., the maximum velocity of nodes v_{max} is zero or close to zero, the geometry relationship among nodes changes slowly. In other words, in each step, little information from network measurements is useful for refining the localization results. In this case, the strategy adopted by [96, 99] is to substitute v_{max} with $v_{max}+\alpha$. As discussed in [96], there is a trade-off with the value of α. On the one hand, the greater the value of α is, the more uncertainty is introduced, because we use a circle with radius $v_{max}+\alpha$ instead of v_{max} to represent the candidate region of each node. On the other hand, if α is small, it cannot provide enough variability in network measurements when the network is static or slowly mobile. Results from [96, 99] demonstrate that setting $\alpha = 0.1r$ works well in practice, where r is the radio range of network nodes.

Chapter 8
Localizability

8.1 Network Localizability

Based on distance ranging techniques, the ground truth of a wireless ad hoc network can be modeled by a distance graph $G = (V, E)$, where V is the set of wireless communication devices (e.g., laptops, RFID tags, or sensor nodes), and there is an un-weighted edge $(i, j) \in E$ if the distance between vertices i and j can be measured or both of them are at known locations, e.g., beacon nodes. Associated with each edge (i, j), we use a function $d(i, j)$: $E \rightarrow R$ to denote the measured distance value between i and j.

An essential question occurs as to whether or not a network is localizable given its distance graph. This is called the *network localizability*. A graph $G = (V, E)$ with possible additional constraints I (such as the known locations of beacon nodes) is *localizable* if there is a unique location $p(v)$ of every node v such that $d(i, j) = \|p(i) - p(j)\|$ for all links (i, j) in E and the constraint I is preserved, where $\|\cdot\|$ denotes the Euclidean distance in the 2D plane. Different from localization that determines locations of wireless nodes, localizability focuses on the location uniqueness of a network.

Localizability assists localization fundamentally and importantly. As previously mentioned, localization often consumes a large amount of computational resource and makes sense only when networks are localizable. Hence, testing localizability before localization can save unnecessary and meaningless computation, as well as accompanying power consumption.

Also, being aware of localizability is of great benefit to many aspects of network operation and management, including topology control, network deployment, mobility control, power scheduling, and geographic routing, as illustrated in Fig. 8.1. Taking deployment adjustment as an example, many measurements (e.g., augmenting communication range, increasing node or beacon density, etc.) can be taken to improve those non-localizable networks to be localizable, which can be effectively guided by the results of localizability testing.

Although the network localizability is given birth by the proliferation of wireless ad-hoc/sensor networks, the problem of unique graph realization has attracted a lot of efforts made by researchers from different literatures over 30 years. An obvious

Y. Liu and Z. Yang, *Location, Localization, and Localizability: Location-awareness Technology for Wireless Networks*, DOI 10.1007/978-1-4419-7371-9_8,

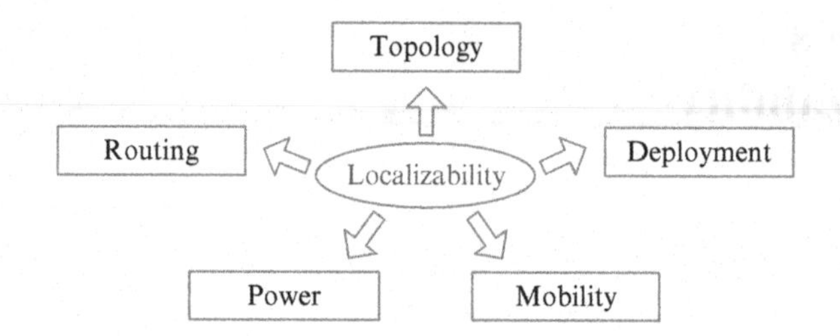

Fig. 8.1 Localizability can assist network operation and management

requirement for a network to be localizable is the network connectivity: when a network is densely connected, it is more likely to be uniquely localizable; otherwise, it may fail the localizability testing. Besides dense connectivity, researchers also point out other requirements for localizable networks, which is discussed in the next section.

8.2 Graph Rigidity

8.2.1 Globally Rigid Graphs

Previous studies have shown that the network localizability problem is closely related to graph rigidity [77, 101–103].

A *realization* of a graph G is a function p that maps the vertices of G to points in Euclidean space (this study assumes 2D space). Generally, realizations are referred to the feasible ones that respect the pairwise distance constraints between a pair of vertices i and j if the edge $(i, j) \in E$. That is, $d(i, j) = \|(p(i) - p(j)\|$ for all $(i, j) \in E$. Two realizations of G are equivalent if they are identical under trivial variation in 2D plane: translations, rotations, and reflections. A distance graph G has at least one feasible realization which represents the ground truth of the corresponding network. Formally, G is embeddable in 2D space and all pairwise distances are compatible, i. e., satisfying the triangle inequality. We assume G is connected and has at least four vertices in the following analysis.

A graph is called *generically rigid* if one cannot continuously deform any of its realizations in the plane while preserving distance constraints [101, 103]. A graph is *generically globally rigid* if it is uniquely realizable under translations, rotations, and reflections. A realization is said to be *generic* if the vertex coordinates are algebraically independent [101]. Since the set of generic realizations is dense in the space of all realizations, we omit this word for simplicity hereafter.

There are several distinct manners in which the nonuniqueness of realization can appear, as shown in Fig. 8.2. A graph that can be continuously deformed while still

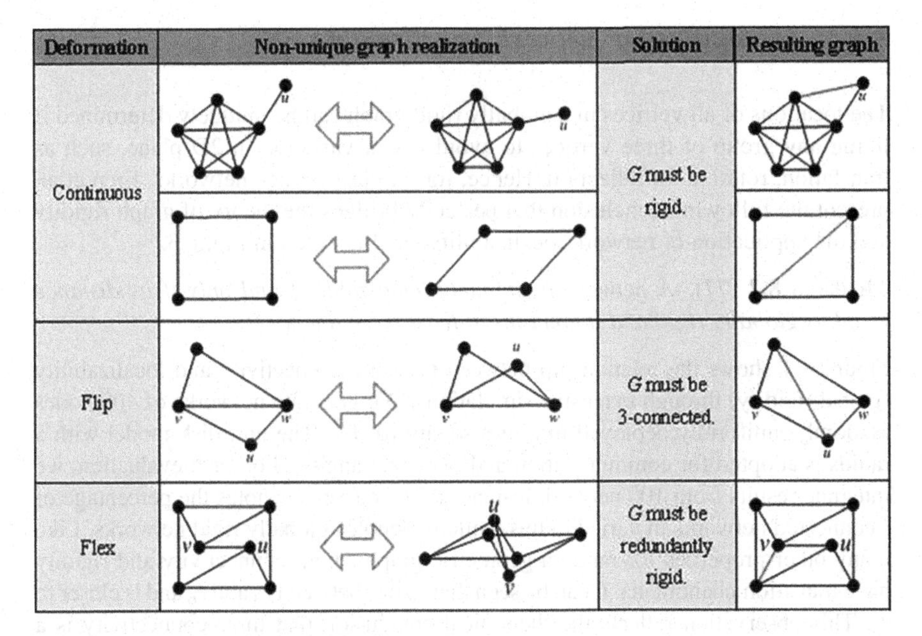

Deformation
Non-unique graph realization
Solution
Resulting graph
Continuous
G must be
rigid.
Flip
G must be
3-connected.
Flex
G must be
redundantly
rigid.

Fig. 8.2 Graph deformation and solutions

satisfying all the constraints is said to be *flexible*; otherwise it is rigid. Hence, rigidity is a necessary condition for global rigidity. Rigid graphs, however, are still susceptible to discontinuous deformation. Specially, they may be subject to *flip* ambiguities in which a set of nodes has two possible configurations corresponding to a "reflection" across a set of mirror nodes (e.g., v and w in the flip example in Fig. 8.2). This type of ambiguity is not possible in three-connected graphs. Figure 8.2 further provides a three-connected and rigid graph which becomes flexible upon removal of an edge. After the removal of the edge (u, v), a subgraph can swing into a different configuration in which the removed edge constraint is satisfied and then reinserted. Such a type of ambiguity, called *flex* deformation, is eliminated by *redundant rigidity*, the property that a graph remains rigid upon removal of any single edge.

Summarizing the conditions for eliminating ambiguities in graph realization, Jackson and Jordan provide the necessary and sufficient condition for global rigidity in Theorem 8.1.

Theorem 8.1 [102]. *A graph with $n \geq 4$ vertices is globally rigid in two dimensions if and only if it is three-connected and redundantly rigid.*

Based on Theorem 8.1, the property of global rigidity can be tested in polynomial time by combining the Pebble game algorithm [104] and the network flow algorithms [101, 105] for rigidity and three-connectivity, respectively.

8.2.2 Conditions for Network Localizability

The locations of all vertices in a globally rigid graph can be uniquely determined if fixing any group of three vertices to avoid trivial variation in 2D plane, such as translation, rotation, or reflection. Hence, for wireless ad hoc networks, Eren et al. present the following conclusion that perfectly bridges the theory of graph rigidity and the application of network localizability, as illustrated in Fig. 8.3.

Theorem 8.2 [77]. *A network is uniquely localizable if and only if its distance graph is globally rigid and it contains at least three anchors.*

Figure 8.4 shows the relationship between network connectivity and localizability (global rigidity) through extensive simulations. We generate networks of 400 nodes randomly, uniformly deployed in a unit square $[0, 1]^2$. The unit disk model with a radius is adopted for communication and distance ranging. For each evaluation, we integrate results from 100 network instances. The curve r_i denotes the percentage of i-connected networks in varied radius while r_g denotes globally rigid networks. Like many other properties for random geometric graphs, both connectivity and rigidity have transition phenomena. It can be seen that r_g lies between r_3 and r_6 and is closer to r_3. This observation reflects the theoretical conclusion that three-connectivity is a necessary condition while six-connectivity is a sufficient one for global rigidity [106].

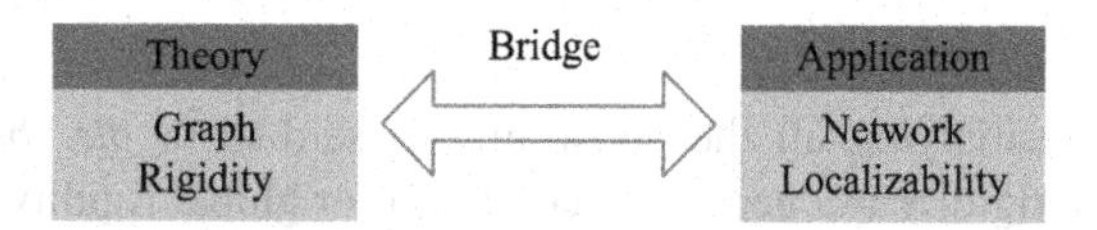

Fig. 8.3 Connection between theory and application

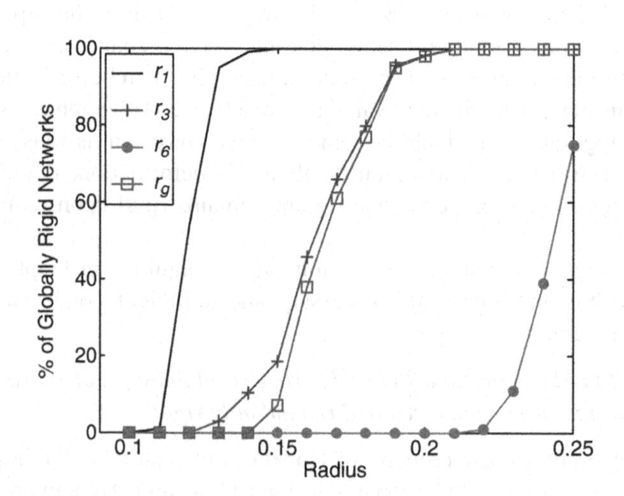

Fig. 8.4 The relationship between connectivity and rigidity

8.3 Inductive Construction of Globally Rigid Graphs

Inductive construction of globally rigid graphs inspires localizability testing in a distributed manner, which is highly appreciated by wireless ad hoc/sensor network community since centralized approaches often consume large communication resource on data transmission and device synchronization.

8.3.1 Trilateration

Trilateration is an important and widely accepted scheme to inductively construct localizable networks. The basic principle of trilateration is that the position of an object can be uniquely determined by measuring the distances to three reference positions. Being employed in many real-world applications [30, 31], it is computationally efficient, fully distributed, and easy to implement. Importantly, the networks that can be constructed by iterative trilateration are localizable.

Theoretically, a *trilateration ordering* of a graph $G = (V, E)$ is an ordering (v_1, $v_2, \ldots, v_n$) of V for which the first three vertices are pairwise connected and at least three edges connect each vertex v_j, $4 \leq j \leq n$, to the set of the first $j-1$ vertices. A graph is a *trilateration extension* if it has a trilateration ordering. It is shown that trilateration extensions are globally rigid [77, 107].

Trilateration-based approaches, however, recognize only a subset (called trilateration extension) of globally rigid graphs. In Fig. 8.5a, two globally rigid

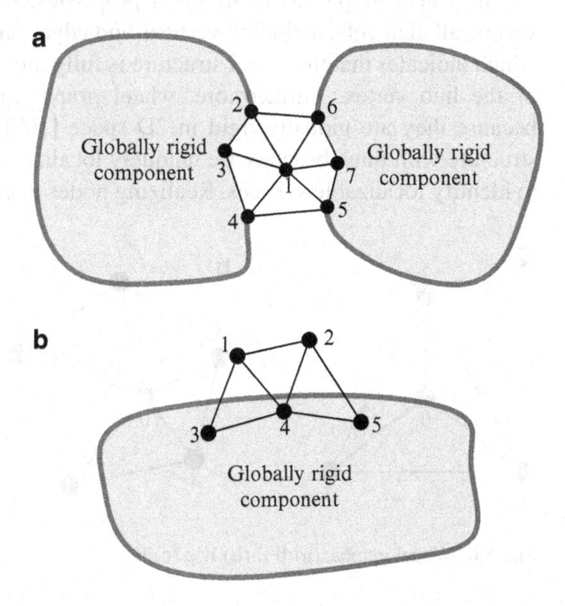

Fig. 8.5 Deficiency of trilateration. (**a**) Geographical map; (**b**) border nodes

components are connected by nodes i ($i = 1,2,\ldots,7$). Suppose the nodes 1, 2, 3, and 4 in the left component are known as localizable. The localizability information, however, cannot propagate to the other part by trilateration since none of the nodes 5, 6, and 7 connects to three localizable nodes. Obviously, trilateration wrongly reports that nodes in the right component are not localizable, ignoring the fact that the entire graph is globally rigid.

A similar situation recurs for the border nodes, as illustrated in Fig. 8.5b. In this case, the border nodes 1 and 2 cannot be localized by trilateration even though nodes 3, 4, and 5 know their locations. Actually, the entire graph in Fig. 8.5b is globally rigid and thus localizable. Discarding locating border nodes is unacceptable, as border nodes often play critical roles in many applications. For example, a sensor network for forbidden region monitoring has special interests on when and where intruders crash into, which are collected by border nodes only.

8.3.2 Wheel

The limitations of trilaterations motivate another method to construct localizable networks based on wheel graphs. A wheel graph W_n is a graph with n vertices, formed by connecting a single vertex to all vertices of an $(n-1)$ cycle. The vertices in the cycle will be referred to as *rim vertices*, the central vertex as the *hub*, an edge between the hub and a rim vertex as a *spoke*, and an edge between two rim vertices as a *rim edge*. Figure 8.6 shows three examples of wheel graphs, in which node 0 is the hub and others are rims.

The wheel graph has many good properties. From the standpoint of the hub vertex, all elements, including vertices and edges, are in its one-hop neighborhood, which indicates that the wheel structure is fully included in the neighborhood graph of the hub vertex. Furthermore, wheel graphs are important for localizability because they are globally rigid in 2D space [108]. Thus, all vertices in a wheel structure with three beacons are uniquely localizable, which indicates an approach to identify localizable vertices. Realizing nodes in general wheel graphs is NP-hard

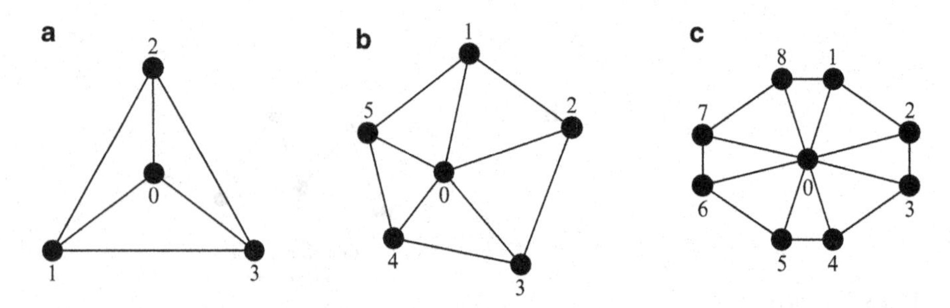

Fig. 8.6 Wheel graphs. (**a**) W_4; (**b**) W_6; (**c**) W_9

[77, 109]. For wireless networks, fortunately, the number d of neighbors of a node (a.k.a., the degree of a node) cannot be arbitrarily high since distance measurements, as well as communication links, only exist between nearby nodes. Therefore, the coordinates of nodes can be calculated by bilateration that examines the location space of at most $O(2^d)$ possible locations, where d is bounded by a constant number.

A wireless network can be modeled by a graph $G_N = (V, E)$. The *closed neighborhood graph* of a vertex $v \in V$, denoted by $N[v]$, is a subgraph of G_N containing only v and its one-hop (direct) neighbors and edges between them in G_N. We also define the *open neighborhood graph* $N(v)$, where $N(v)$ is obtained by removing v and all edges incident to v from $N[v]$. Note that $N[v]$ is the local information known by the vertex v.

According to the previous analysis, if a vertex in $N[v]$ is included in a wheel graph centered at v, it is localizable by given three beacons in $N[v]$. The localizability issue now can be transformed to finding wheel vertices in $N[v]$ when given a number of known localizable vertices.

We first consider the presence of three localizable vertices in $N[v]$. There are two cases of their distribution: (1) the hub v and two rim vertices and (2), three rim vertices. In the second case, v can be easily localized by trilateration. As a result, this case degenerates to the first one. We thus focus on the first case in the following analysis. Without loss of generality, suppose the two rim localizable vertices are v_1 and v_2. In addition, a two-connected component in a graph G is a maximal subgraph of G without any *articulation* vertex whose removal will disconnect G. For simplicity, we use *blocks* to denote two-connected components henceforth if no confusion caused.

In Theorem 8.3, we propose a sufficient and necessary condition to find wheel vertices.

Theorem 8.3 [108]. *In a neighborhood graph $N[v]$ with k ($k \geq 3$) localizable vertices v_i ($i=1, \ldots, k$ and $v = v_k$), any vertex (other than v_i) belongs to a wheel structure with at least three localizable vertices if and only if it is included in the only block of $N(v)$ that contains $k-1$ localizable vertices.*

According to Theorem 8.3, finding wheel vertices can be implemented by calculating blocks, as shown in Algorithm 8.1. Suppose there are k localizable vertices in a neighborhood graph $N[v]$.

The core part of Algorithm 8.1 is to find blocks in a graph $G=(V, E)$. This can be done by depth first search in linear time in terms of the size of graphs. Hence the time complexity of Algorithm 8.1 is $O(|V|+|E|)$.

Theorem 8.4 (Correctness) [108]. *In a neighborhood graph $N[v]$, a vertex is marked by Algorithm* 8.1 *if and only if it is uniquely localizable in $N[v]$.*

Theorem 8.4 guarantees the optimality of Algorithm 8.1 since it finds the maximum number of localizable vertices in $N[v]$.

Now, we consider the localizability for an entire network. We call this problem the network-wide localizability test so as to distinguish with the previously

Algorithm 8.1 Node localizability

```
1:  if k>=3, then
2:    find all blocks in N(v), denoted by B_i, i=1,...,m;
      let B_1 be the unique one of localizable nodes;
3:    for each vertex x not being marked in B_1
4:      mark x localizable;
5:      connect x to all other localizable ones;
6:    end for
7:  end if
```

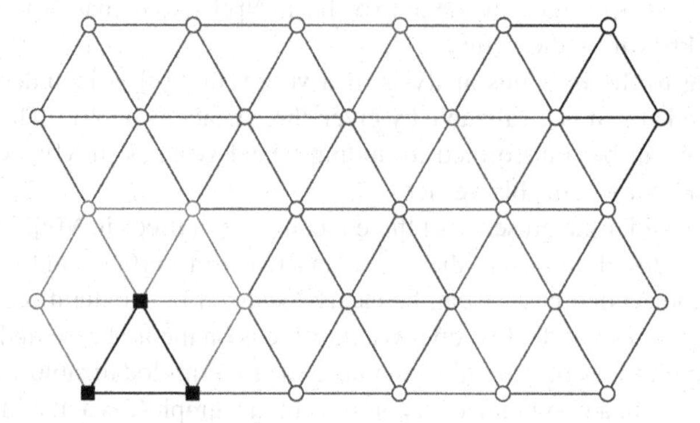

Fig. 8.7 A wheel extension graph

discussed case of localizability within neighborhood. Similar to the trilateration extension, the wheel extension is defined as follows.

Definition 8.1. *A graph G is a wheel extension if there are*

(a) *three pairwise connected vertices, say v_1, v_2, and v_3; and*
(b) *an ordering of remaining vertices as v_4, v_5, v_6..., such that any v_i is included in a wheel graph (a subgraph of G) containing three early vertices in the sequence.*

Theorem 8.5 [108]. *The wheel extension is globally rigid.*

The family of wheel extensions is actually a superset of trilateration extensions. Figure 8.7 shows an example that is a wheel extension but not a trilateration extension. To test localizability, it is important to know whether a graph is a wheel extension. In the following, we present a distributed protocol that can mark localizable nodes in a network. The protocol works in an iterative manner in which a node marked in the current iteration acts as a known localizable one (or beacon) in subsequent iterations. Localizability information diffuses step by step and reaches the entire network after a number of iterations. A particular iterative process on the example graph shown in Fig. 8.7 is as follows. First, three beacons are available at the bottom left. In the first iteration, nodes in the bottom left hexagon are identified because they are included in a wheel graph with three beacons. Such a procedure continues until all localizable nodes are marked.

The protocol is given in Algorithm 8.2, which is conducted in a distributed manner at each node. If all nodes in a network are marked by Algorithm 8.2, the network graph is a wheel extension, and vice versa.

We now analyze the time complexity of Algorithm 8.2 running on a graph G with n vertices. Since Algorithm 8.1 is only executed at the vertices with at least three localizable ones in $N[v]$, these vertices are localizable and will be marked by Algorithm 8.2. Therefore, the running time of Algorithm 8.2 is output sensitive. In the worst case, Algorithm 8.1 will be executed in all vertices in G. Let $d(v)$ denote the degree of a vertex v. In line 2, calculating blocks in $N(v)$ costs $O(d(v)^2)$ time in dense graphs or $O(d(v))$ in sparse graphs. In the while loop between lines 3 and 11, at most $d(v)$ neighbors are marked and informed. Hence, the total running time of Algorithm 8.2 is $\sum_{v \in G} O(d(v)^2 + d(v)) = O(n^3)$ in dense graphs and $\sum_{v \in G} O(d(v)) = O(n)$ in sparse graphs. The bound is tight due to the instance of $G = K_n$, where K_n is the complete graph of n vertices.

In practice, a wireless ad hoc network cannot be excessively dense because the communication links only exist between nearby nodes due to radio signal attenuation. In addition, the mechanism of topology control reduces redundant links to alleviate collision and interference. Hence, the proposed algorithm is practically efficient.

To analyze the correctness of Algorithm 8.2, we first define the concept of k-hop localizability.

Definition 8.2. *In a network, a node is* k-hop localizable if it can be localized by using only the information of at most k-hop neighbors.

Algorithm 8.2 Network localizability

```
1:   exchange neighbor list between neighbors;
2:   construct N[v];
3:   if N[v] has >= 3 localizable nodes
4:      run Algorithm 8:1 on N(v), obtaining a number of blocks B_i; (Assume B_1 is the
        unique localizable one)
5:      mark v and B_1 localizable;
6:      inform B_1 the change;
7:      Update N(v);
8:   end if;
9:   while(true)
10:     wait for state change of neighbor nodes;
11:     update N(v);
12:     if any nonmarked B_i has >=2 localizable nodes
13:        mark B_i localizable;
14:        update N(v);
15:        inform Bi the change;
16:     end if
17:  end while
```

Clearly, 1-hop localizable is the most critical condition for all k and the set of k-hop localizable nodes is monotonically increasing.

Theorem 8.6. *In a graph G, a vertex marked by Algorithm* 8.2 *if and only if it is 1-hop localizable in G.*

Theorem 8.6 suggests the optimality of Algorithm 8.2 by showing that it is able to recognize all 1-hop localizable nodes locally.

Compared to the previous trilateration (TRI)-based methods, the advantages of the proposed method (WHEEL) lie in:

1. *Capability*. Recognizing a superset of localizable nodes, as shown in Fig. 8.8.
2. *Efficiency*. Taking $O(n)$ running time for sparse graphs and $O(n^3)$ for dense ones, where n is the network size.
3. *Low cost*. Introducing no extra wireless communication cost by using only localized information.

Three examples are further provided to show how wheel outperforms TRI. In Fig. 8.9, a particular network with an "H" hole is generated in which 400 nodes are randomly distributed. The blue dots denote the nodes marked by TRI, while reds denote the nodes marked by wheel but not by TRI. Neither TRI nor wheel can mark the remaining blacks. Wheel can easily step over gaps, such as borders or barriers, and recognize more nodes than TRI does. The same phenomenon recurs in all three network instances.

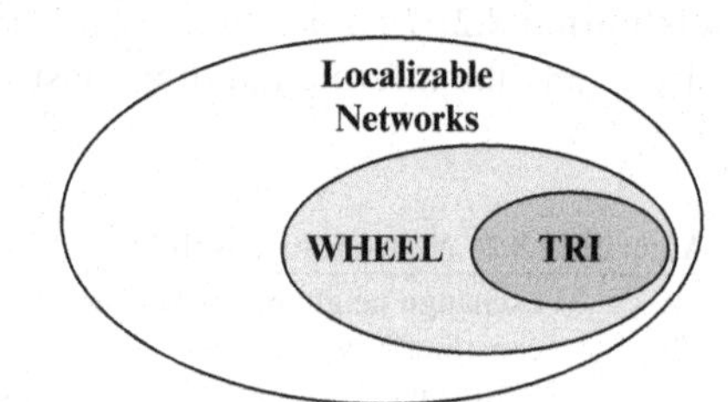

Fig. 8.8 Trilateration extension is a subset of wheel extension

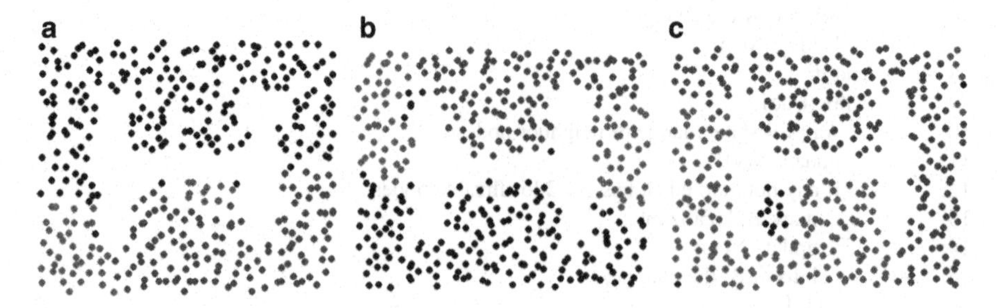

Fig. 8.9 Networks with "H" holes. (**a**) Case 1; (**b**) case 2; (**c**) case 3

8.4 Node Localizability

Due to hardware or deployment constraints, for some applications, the networks are almost always not entirely localizable [110]. Indeed, theoretical analyses indicate that, in most cases, it is unlikely that all nodes in a network are localizable, but a (large) portion of nodes can be uniquely located [110]. Thus, the network localizability testing often fails unless networks are highly dense and regular.

On the other hand, nodes are not equally important since they play different roles in a network. Such differentiation can be application specific. For example, a sensor network for monitoring forbidden regions has special interest in when and where the intruders enter, which are collected by border nodes only. In addition, many applications can function properly as long as a sufficient number of nodes are aware of their locations [110]. These observations motivate researchers to consider the localizability problem beyond the network localizability.

Although the theory of network localizability is complete, what we really desire is to answer the following two fundamental questions that cannot be solved by existing methods:

1. Given a network configuration, whether or not a specific node is localizable?
2. How many nodes in a network can be located and what are they?

Answering the above questions not only benefits localization, but also provides instructive directions to other location-based services. Therefore, the *node localizability* is addressed [110, 111], which focuses on the location uniqueness of every single node. Clearly, network localizability is a special case of node localizability in which all nodes are localizable. Thus, node localizability is a more general issue.

The first major challenge for studying node localizability is to identify uniquely localizable nodes. Following the results for network localizability, an obvious solution is to find a localizable subgraph from the distance graph and identify all the nodes in the subgraph localizable. Unfortunately, such a straightforward attempt misses some localizable nodes and wrongly identifies them as nonlocalizable, since some conditions (e.g., three-connectivity) essential to network localizability are no longer necessary to node localizability. As shown in Fig. 8.10a, node u can be uniquely located under this network configuration but not included in the three-connected component of beacons. The uniqueness of u's location is explained in Fig. 8.10b, c where we decompose the network into two subgraphs. As u connects two beacons in the right component, it has two possible locations denoted by u and u'. If we adopt u' as its location, it is impossible to embed the left subgraph into the plane. Specifically, the left subgraph has two realizations, but neither of them is compatible with u'. Hence, u is uniquely localizable, although the three-connectivity property does not hold. Motivated by the example, it is clear that the results derived for network localizability cannot be directly applied, and we have to reconsider the conditions for node localizability.

The big picture of the state of the arts of node localizability is shown in Fig. 8.11. An obvious condition for a vertex to be localizable is that it must connect at least

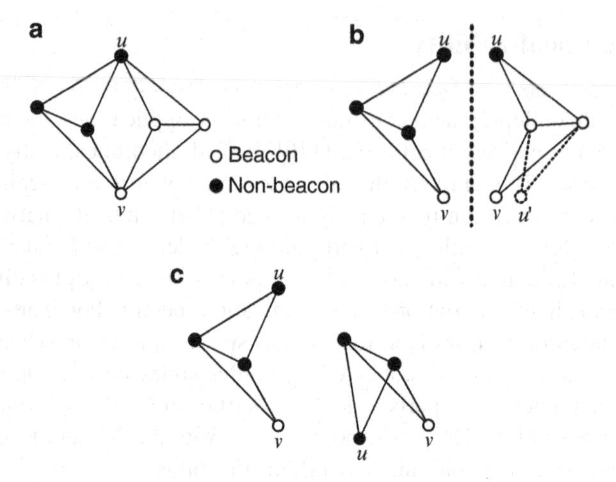

Fig. 8.10 An example showing that the result from network localizability fails to identify node u as localizable. (**a**) u is localizable; (**b**) graph decomposing; (**c**) two realizations of the left subgraph

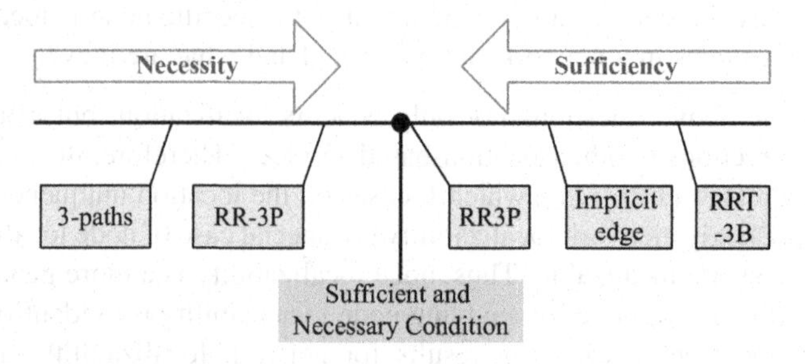

Fig. 8.11 Evolution of conditions of node localizability

three other nodes, i.e., having a degree of at least 3. Goldenberg et al. [110] propose the first nontrivial necessary condition: if a vertex is localizable, it has three vertex-disjoint paths to three beacons. We denote such a condition as three-paths for short. The necessity of three-paths is obvious: if a vertex has only two vertex-disjoint paths to beacons, it definitely suffers from a potential flip ambiguity by reflecting along the line of a pair of cut vertices. Nevertheless, it is easy to find that some nonlocalizable vertices also satisfy the three-path condition, as illustrated in Fig. 8.12. The vertex u is flexible even though it has three vertex-disjoint paths to three beacons. Redundant rigidity has been further proved to be essential for node localizability and accordingly Yang et al. [111] achieve the best necessary condition by combining 3P (three vertex-disjoint paths) and redundant rigidity, which is called RR–3P for short. Clearly, RR–3P is still insufficient as illustrated in Fig. 8.13a. Considering the vertex u, it satisfies the RR–3P condition but not localizable due to the discontinuous flexing in which u can reflect along the axis denoted by the dashed line in Fig. 8.13b.

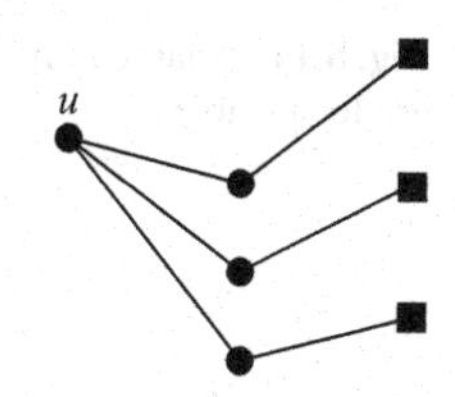

Fig. 8.12 The condition of three paths is insufficient

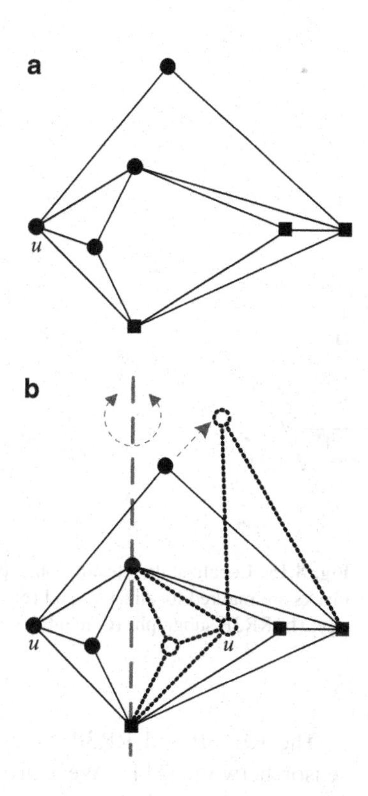

Fig. 8.13 RR–3P is insufficient. (**a**) The vertex u satisfies RR–3P; (**b**) u suffers a discontinuous flexing

On the other hand, for the sufficiency of node localizability, a straightforward solution [110] is presented to identify localizable nodes by finding globally rigid subgraph. All nodes in a globally rigid subgraph with at least three beacons (denoted by RRT-3B) are localizable. An improvement has also been made by introducing the "implicit edge." Recent studies show that a vertex is localizable if it belongs to a redundantly rigid component in which there exist three vertex-disjoint paths connecting it to three beacon vertices in the component (denoted by RR3P) [111]. Note that RR3P is fundamentally different from the previously mentioned RR–3P. RR3P requires the three paths strictly residing in the redundantly rigid component.

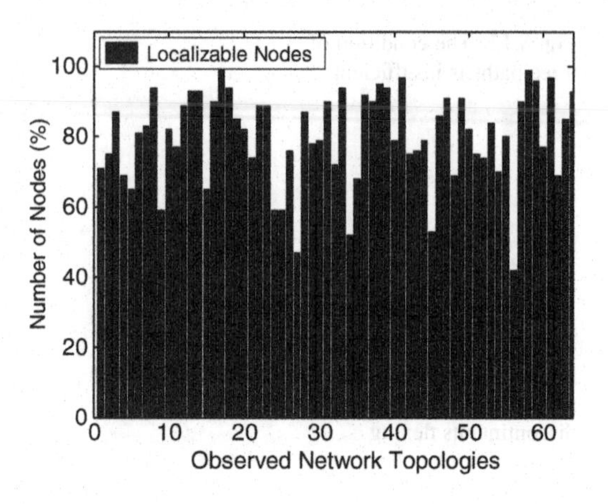

Fig. 8.14 A large portion of nodes are localizable

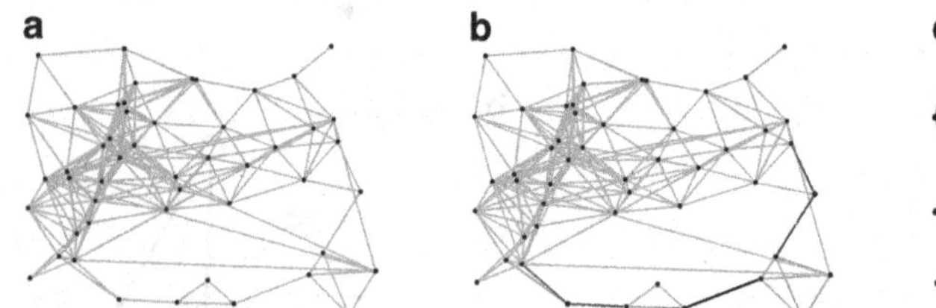

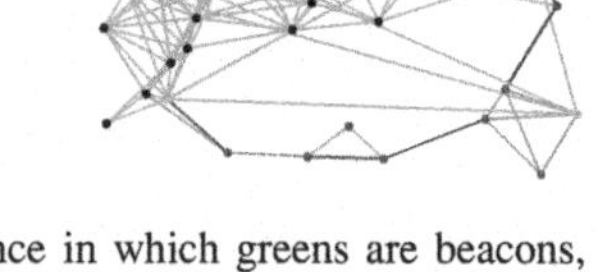

Fig. 8.15 Localizability testing on a particular network instance in which greens are beacons, blacks are marked localizable, and reds are marked nonlocalizable. (**a**) Measured network topology; (**b**) RR3P subgraph; (**c**) identifying localizable nodes

The RR–3P and RR3P conditions are implemented in a real-world wireless sensor network [111]. We can see from Fig. 8.14 that almost all the time the network is not entirely localizable. However, a large portion, on average nearly 80%, of nodes are actually localizable (i.e., identified by the RR3P condition). Specifically, 90% of network topologies have at least 60% of the nodes localizable, and more than 25% of topologies have at least 90% of nodes localizable. These results suggest the necessity and importance of node localizability. Figure 8.15 shows the results of node localizability testing. For the first time, it is possible to analyze how many nodes one can expect to locate in sparsely or moderately connected networks.

Other than figuring out localizable nodes, being aware of node localizability greatly helps network deployments. Generally speaking, for those nonlocalizable networks, we expect to make them localizable by adjusting some network parameters. Traditional solutions include augmenting ranging capability, increasing node density, or equipping more nodes with GPS. Such measures can be more

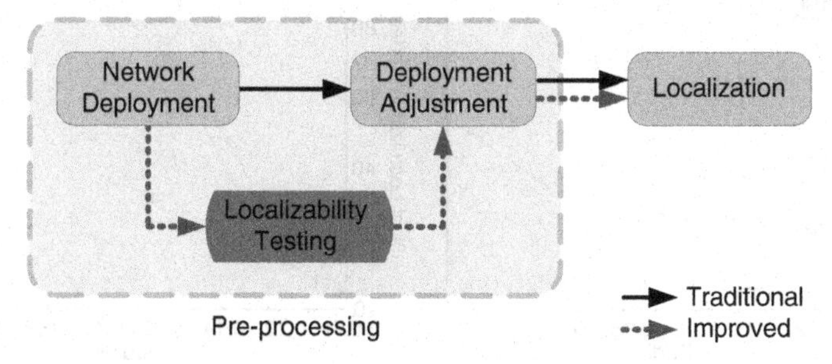

Fig. 8.16 Localizability assists network deployment

targeted and effective with the knowledge of node localizability. Specifically, it is possible for these adjustments to focus on nonlocalizable nodes only instead of blindly exerting on all nodes.

Similar to existing localization approaches, the improved localization approach can be divided into two stages: data preprocessing and location computation. As a rule, the deployment adjustment is included in the preprocessing stage so as to intensify network localizability or reduce the computation complexity of localization. As shown in Fig. 8.16, the major difference of the improved flow is that the task of localizability testing is added to assist deployment adjustment. Specifically, the testing algorithm is carried out on the initial network deployment and the results are used to instruct the subsequent adjustments.

We increase the distance ranging capability by augmenting signal transmitting power. Those localizable nodes keep their states unchanged while others augment their distance ranging capabilities to proper levels. This network adjustment not only increases the number of localizable nodes, but also decreases communication interference and energy consumption compared with traditional solutions.

Large-scale simulations are further conducted to examine RR–3P and RR3P under varied network parameters. We randomly generate networks of 400 nodes, uniformly deployed in a unit square $[0, 1]^2$. The unit disk model with a radius is adopted for communication and distance ranging. For each evaluation, we integrate results from 100 network instances.

We study the improvements of our proposed conditions to existing ones. Note that the necessary conditions and the sufficient ones can be used to identify nonlocalizable and localizable nodes in a network, respectively. Other than the proposed RR–3P and RR3P, for comparison, we introduce the best previous necessary condition 3P and the widely used sufficient condition TRI, which is the theoretical upper bound of trilateration based approaches. Figure 8.17a shows the amount of nodes marked by 3P and TRI. As we know, nodes above the curve of 3P are nonlocalizable while those below the curve of TRI are localizable. In addition, the other ones between two curves are unknown at present based on 3P and TRI. Specifically, almost 70% of nodes left unknown at radius 0.18. Contrastively, Fig. 8.17b shows the results if we adopt the proposed RR–3P and RR3P. Clearly, two curves are close to each other and the gap between them is always narrow along

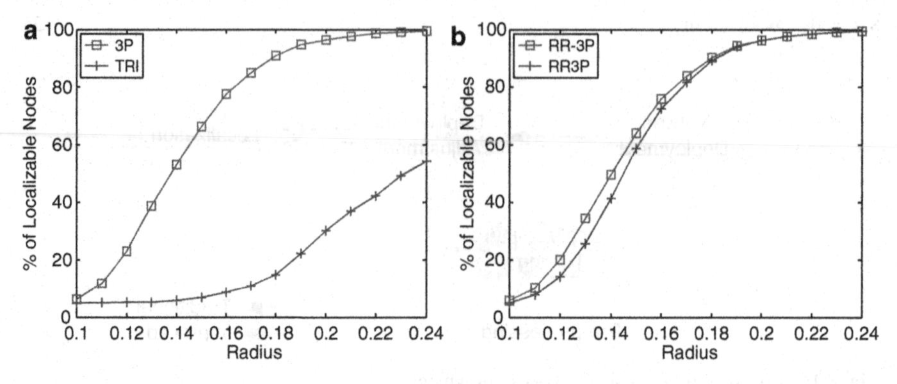

Fig. 8.17 Improvements of proposed RR–3P and RR3P. (**a**) The capability of 3P and TRI; (**b**) the capability of RR–3P and RR3P

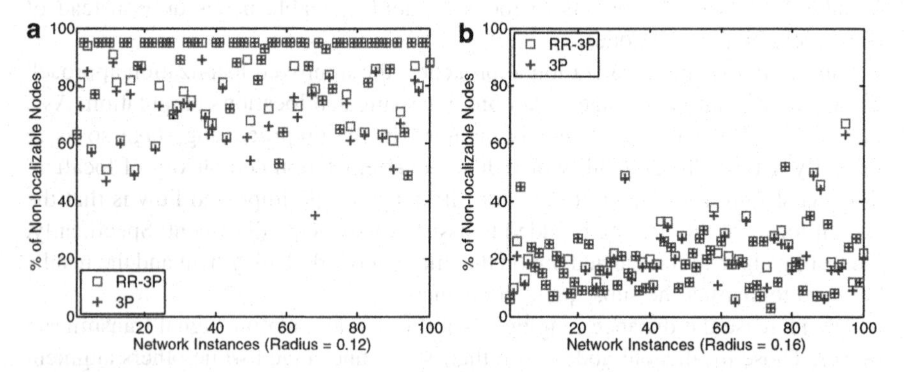

Fig. 8.18 Comparison between necessary conditions: 3P and RR–3P. (**a**) Sparse; (**b**) Medium

with the variation of network connectivity, indicating a smaller number of nodes whose localizability cannot be determined.

We also study the performance of node localizability for sparsely and moderately connected networks. In this evaluation, the percentage of localizable and nonlocalizable nodes in 100 network instances is shown in Figs. 8.18 and 8.19 with communication radius r=0.12 and 0.16. According to Fig. 8.18a, b, RR–3P and 3P have nearly similar capabilities to recognize nonlocalizable nodes at both sparse and medium network connectivity, except for a few cases in which RR–3P successes much. For sufficient conditions, as shown in Fig. 8.19a, RR3P identifies 30% nodes as localizable while TRI cannot work at all due to sparseness. When $r = 0.16$ in Fig. 8.19b, RR3P recognizes, on average, more than 70% localizable nodes in 78 cases while TRI only marks less than 10% localizable ones in 91 cases. Such observations show that RR3P remarkably outperforms TRI at a specific range of communication radius.

We further provide two examples to show how RR–3P and RR3P outperform 3P and TRI. In Fig. 8.20, a particular network with a "Z" hole is generated in which 400 nodes are randomly distributed. The red dots denote the localizable nodes marked by TRI while blues denote the nonlocalizable nodes marked by 3P. Neither TRI nor 3P can recognize the remaining gray ones. As shown in Fig. 8.21, similar evaluations are conducted on the same data sets and we use RR3P and RR–3P

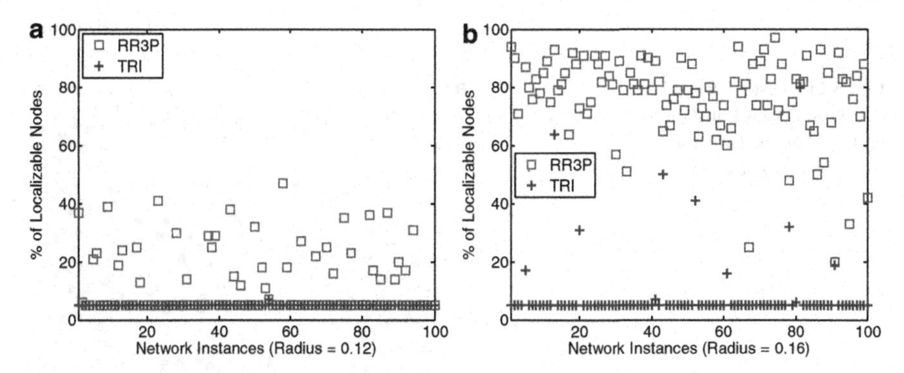

Fig. 8.19 Comparison between sufficient conditions: TRI and RR3P. (**a**) Sparse; (**b**) Medium

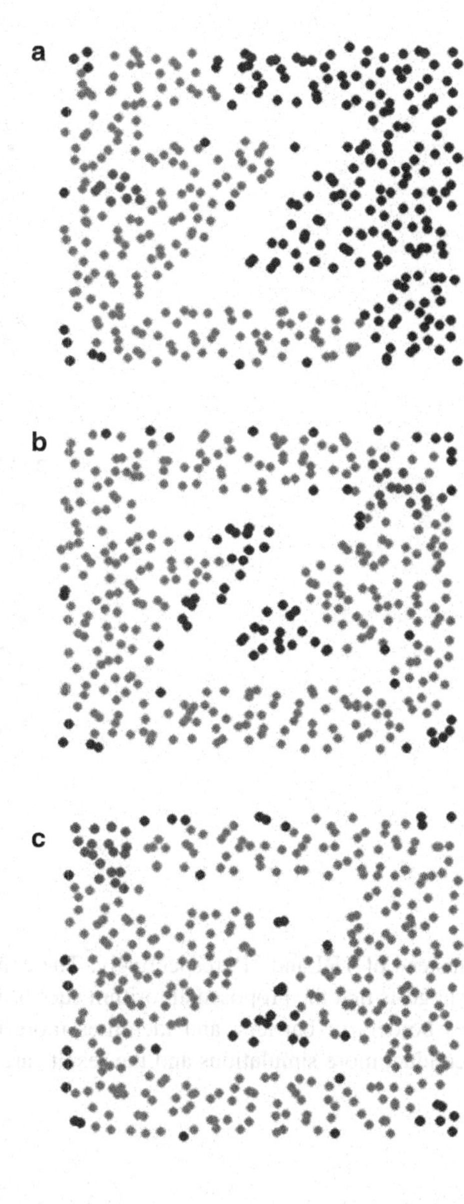

Fig. 8.20 Testing 3P and TRI on network instances with "Z" holes. (**a**) Case 1; (**b**) case 2; (**c**) case 3

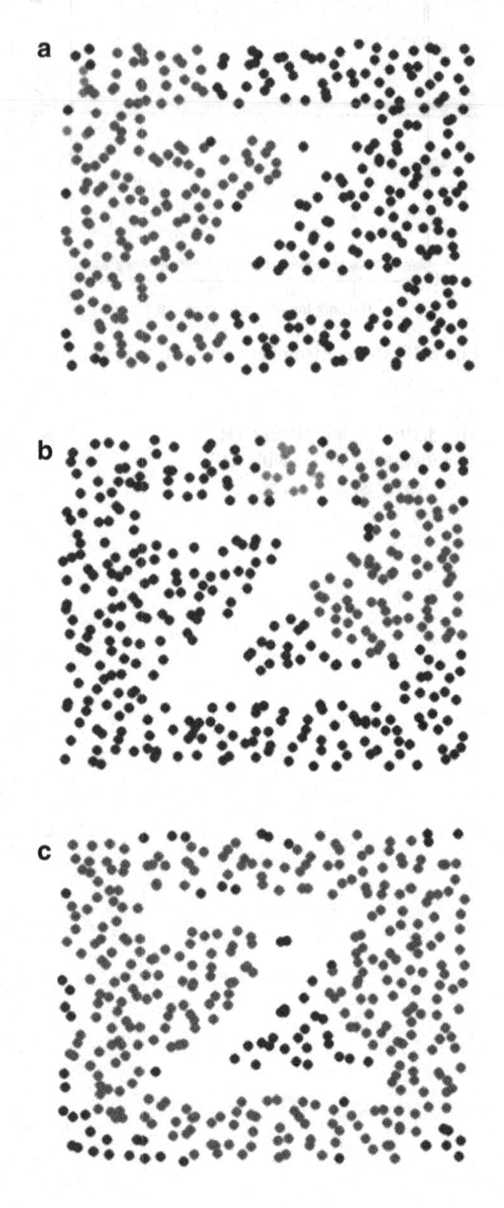

Fig. 8.21 Testing RR–3P and RR3P on network instances with "Z" holes. (**a**) Case 1; (**b**) case 2; (**c**) case 3

instead of TRI and 3P, respectively. The comparison between Figs. 8.20 and 8.21 suggests that the proposed algorithm successfully steps over geographic gaps, such as borders or barriers, and identifies more nodes than previous approaches. We conduct more simulations and the results are consistent.

8.5 Summary

In this chapter, we discuss the localizability issue for wireless networks. Network localizability is to determine whether or not all nodes in a network are localizable given distance constrains. In recent years, this issue draws remarkable attentions from an increasing number of researchers. Based on rigidity theory, we analyze the reasons why the locations of nodes in a network cannot be uniquely determined. In addition, we present two approaches for inductively constructing globally rigid graphs: trilateration and WHEEL. WHEEL is proved to be a nice substitute for trilateration, determining the locations of a larger number of nodes, more suitable for sparsely or moderately connected networks, and introducing no extra communication cost. We then introduce the concept of node localizability. Different from network localizability, node localizability focuses on whether or not a specific node is localizable given distance constrains. Node localizability is a more general issue and accordingly more difficult. We extensively investigate the state-of-the-art results on finding the condition for a node being localizable. In general, this is a new research area, leaving several important problems unsolved as follows.

Currently, the necessary and sufficient condition for node localizability is still open and researchers believe RR3P is the one. The sufficiency of RR3P is proved in [111]; while the necessity remains open, which is both challenging and worthwhile.

Another direction of future research with good potential is localizability under noisy distance measurements. Previous studies have shown that measurement errors play an important role on localization. Some nodes uniquely localizable under perfect distance ranging may suffer from location ambiguities in a practical scenario of ranging errors. We envision this point in order to increase the robustness of localizability testing.

Chapter 9
Location Privacy

9.1 Introduction

Location-based services exploit location information to provide a variety of fancy applications. For instance, E-911 in the USA (correspondingly E-112 in Europe) tries to help the caller of an emergency call as soon as possible by locating him or her through GPS. Besides, applications that notify users the nearby places of interest (such as the nearest hospital, restaurant, and store.) can facilitate daily life. While amount of attractive quality of life enhancing applications are presented by location-based services, new threats are also brought in. Among these threats, perhaps the most important one is the intrusion of location privacy.

To clarify the meaning of the term "location privacy," we use Alan Westin's commonly quoted definitions of information privacy [112]. Location privacy can be defined as a special type of information privacy:

> *Location privacy is the claim of individuals, groups, or institutions to determine for themselves when, how, and to what extent location information about them is communicated to others.*

In other words, location privacy mainly concerns user's ability of controlling location information.

Most people have not paid enough attention on their location privacy. They tend to underestimate the harm of location leaking for possible two reasons. First, they do not fully understand the negative consequence of privacy divulging. Along with the proliferation of pervasive and mobile computing, however, location disclosure not only leaks location information, but also leads to the implications of personal information. For example, by tracking the history of one's movement, it is possible for attackers to reveal some personal information, such as who he is, where he usually goes shopping, what company he is working for, and how often he does exercise.

Second, protecting location privacy usually to some extent sacrifices the quality of services. Therefore, when we study location privacy, there is a key question throughout: How much protection on location privacy is effective and acceptable? Although the answer to this question is actually application and user dependent, the public has a common belief that a good service design should take both the quality and the privacy concerns into account.

Y. Liu and Z. Yang, *Location, Localization, and Localizability: Location-awareness Technology for Wireless Networks*, DOI 10.1007/978-1-4419-7371-9_9,

In Sect. 9.2, we discuss the threats on location privacy. Section 9.3 discusses the four classes of privacy protection strategies. In Sect. 9.4, we concentrate on location anonymity, which involves most of recent research works. Section 9.5 provides several directions of ongoing research on location privacy.

9.2 Threats

To illustrate the threats of location privacy, we focus on two questions: How can adversary obtain the location information of others? What if location information is leaked?

9.2.1 How Can the Adversary Obtain Location Information of Others?

Most users only desire to release their location information to certain service providers. A straightforward question is how a third-party adversary can get access to the location information.

There are several possible ways. For example, an adversary can intercept the communications between the user and the service provider, or crack data from the service provider directly, if the service provider does not protect user data well. What is worse, some service providers might be camouflaged and malicious, so they intentionally collect user information and sell them to hostile parties.

9.2.2 What Is the Negative Consequence of a Location Leak?

The second question is what the consequence of location leaking is. A direct negative effect is that personal well-being and safety are influenced. The leakage of location information not only yields the uncomfortable creepiness of being watched, but also leads to physical harms to individuals.

Another negative effect is the unwanted revelation of user activities. For most people, it might be embarrassing to be seen at places such as abortion clinic and AIDS clinic. It might also be unwilling for a staff if the proximity to a business competitor is revealed to the boss. Generally speaking, location information consists of three explicit or implicit factors: time, location, and personal identity. Therefore, a large amount of personal information, such as political affiliations, religious beliefs, lifestyles, and medical status, can be inferred by gathering location information.

Here we use the term "gathering" because rather than the presence at certain locations, the pattern of movement can be acquired by tracking an individual's location for a period and help the adversary to read the meaning of the individual's routes.

The following example shows how one's home address can be inferred from one's pattern of everyday movement. Assume his location is recorded by an attacker every 5 min. Then all these location information can be segmented into discrete trips. Observing these trips long enough (say, at least 1 km long), the adversary can gather many clues in order to infer the location of his home. First, if the last trip always ends in a same place everyday, this place has a high probability of being his home. Second, if the subject spends much more time in a same place than in other places, then this place may be his home. Third, considering the place of his stay between 6 p.m. and 8 a.m., if there is a place that occupies a high percentage, that place is probably his home.

9.3 Protection Strategies

In [113], existing location privacy protection strategies fall into four categories: regulatory, privacy policies, anonymity, and obfuscation. Regulatory strategies try to govern the use of personal information by legislation. Privacy policies provide flexible privacy protection in order to meet the different requirement of users. Anonymity approaches aim at disassociating the location information from the real identity of a user. Obfuscation protects privacy by degrading the resolution of location information provided by service providers. The former two strategies mainly aim at preventing the attacker from obtaining the location information of others through political efforts of mechanism designs. The latter two, on the contrary, aim to preserve location privacy technically.

9.3.1 Regulatory Approaches

The most fundamental privacy protection strategy is to govern fair use of personal location information by developing related regulations. Existing regulations are quite different from one another since they are drafted by different organizations and nations based on their own requirements. These regulations can be mainly summarized by the five core principles proposed in Fair Information Practice Principles [114]:

1. *Notice/awareness*. Individuals must be aware of the identification of the entity collecting the data and the purpose of data collecting.
2. *Choice/consent*. Individuals must be able to decide how any personal information collected from them may be used.
3. *Access/participation*. Individuals must be able to access data about themselves and to contest the data accuracy and completeness.

4. *Integrity/security*. Collectors must ensure the accuracy of personal data and protect these data from disclosure.
5. *Enforcement/redress*. Collectors must be accountable for any violation of above principles.

Although legislation provides a powerful way of protecting privacy, it also brings about troubles. Privacy laws vary from nation to nation, so that location-based services abide by the laws of a particular nation might violate privacy rules of another nation. This issue makes it difficult for service providers to extend their business in different nations without changing the services.

Another issue is that regulations only ensure the mechanisms of enforcement and accountability when a violation of location privacy is detected. They cannot prevent invasions of privacy afore. Moreover, regulation legislating always lags behind the development of new technologies.

9.3.2 Privacy Policies

Regulation provides global or group-based protection of privacy, while it lacks flexibility. Different individuals may have different concerns about their location privacy. A super star might be very sensitive about the disclosure of his location, but for ordinary people, most of their location information is less interesting to the public.

Privacy policies aim at providing flexible privacy protection by adopting individual requirements. They are trust-based mechanisms. The term "trust based" means that the system must be trusted by the users. Policy-based approaches cannot provide privacy if the system betrays.

PIDF (presence information data format) [115] is a location privacy policy scheme adopted by the IETF (Internet engineering task force). A user specifies his acceptable usage of location information, such as whether retransmission of the data is allowed, at what time the data expire and should be discarded, etc. Personal preference of privacy policy is then attached to the location information to be submitted. Both location information and privacy policy are encapsulated into a location object and digitally signed (in order to prevent separating the location information from privacy policy) before sending out.

P3P (privacy preferences project) [116] is a Web-based privacy protection mechanism developed by W3C (World-Wide Web consortium). Unlike PIDF, P3P focuses on the service providers rather than the users. Service providers can publish their data practices, including the purpose of data collecting, how long would these data be held, and whom might these data be shared with. And it leaves for the users proscribing a particular service to decide whether its data practices violate their own privacy requirement. P3P does not explicitly address location privacy issues, while its mechanism can be extended for location awareness context.

There are other policy-based mechanisms for location privacy protection, such as PDRM (personal digital rights management) [117] and IBM's EPAL (enterprise privacy authorization language) [118]. All these policy-based initiatives only provide a partial solution to privacy. The practicality of these policies under location-aware environment, which involves frequent and dynamic location information, is not yet proved. Unlike the regulatory approaches, privacy policies provide no enforcement, but rely on economic, social, and regulatory pressures.

9.3.3 Anonymity

As mentioned in Sect. 9.2, adversary inference mainly counts on the three factors: time, location, and personal identity. A direct thought is that if we can hide the personal identity, i.e., make the released location information anonymous, we can avoid being affected by the disclosure of location information, because even some inferences are successfully obtained, an attacker still has no idea about the identity of the subject.

Anonymity is a technical countermeasure that dissociates information about an individual from his identity. Its goal is to use location-based services without revealing user identity. Unlike the trust-based mechanisms, anonymity-based approaches always suspect every service provider. A service intermediary is introduced for anonymity-based scheme, which is trusted and might help users hiding their identities. In such a scheme, users do not communicate with service providers directly. Instead, they communicate with the intermediary first, and then the intermediary would fetch data from the service providers and send the data back to the users. The design of a service intermediary is important for both service providers and users.

Notice that, it is clear that some location-based services, such as "when I am at home, let my family know where I am" cannot work without the identity of the user. The anonymity-based approaches mainly focus on other types of services that can work in the absence of real identities, such as "when I walk into a restaurant, show me the menu." In Sect. 9.4, we discuss anonymity-based approaches in detail.

There are drawbacks for anonymity-based approaches. First, anonymity-based approaches usually rely on the design and deployment of the intermediary. Second, anonymity barriers authentication and personalization, and thus prevents some customized applications.

9.3.4 Obfuscation

Obfuscation deliberately degrades the resolution of location information in order to protect privacy while allowing user identities to be revealed. There are three types of imperfection in the literature that can be introduced into the location information: *inaccuracy*, *imprecision*, and *vagueness*. In location awareness context, inaccuracy means telling a location differs from the real location; imprecision means telling a region including the real location instead of the real location; and vagueness means involving linguistic terms like "near" or "far from" in the conveyed location. Many researches on obfuscation concentrate on the use of imprecision.

Some anonymity-based approaches also use imprecision. The difference of anonymity and obfuscation is that anonymity aims to make an individual indiscernible to a number of other individuals, while obfuscation aims to make the location of an individual indiscernible to a number of other locations.

Commonly used in location-based services, proximity query typically asks about the life facilities close to a user's location, e.g., "where is the nearest restaurant?". In [119], an algorithmic approach is proposed to obfuscating proximity queries. An individual reports a set O of locations instead of his real location. The service provider then tries to find the position of interest for each location in O. If all locations in O have the same result, the provider can return this result to the user. Otherwise, it asks whether the user agrees to refine his location. If the user agrees to do so, the algorithm reiterates. If the user refuses, the provider returns the best estimate approximation according to the coarse-grained information provided by the user.

Obfuscation does not rely on any intermediary, and users can communicate with service providers directly. As a result, the architecture is lightweight and distributed. Also, it enables the applications that require authentication or personalization, which might be blocked for the anonymity-based approaches. Even though researchers claim that most location-based services can work with imprecise location, the loss of quality of service is left open for study.

9.4 Anonymity-Based Approaches

Releasing location information anonymously (i.e., using a pseudonym instead of an actual identity) can prevent attackers from linking the location information to an individual. However, hiding the name is not enough. It is possible for attackers to reidentify an individual from the location information of a pseudonym. For example, certain regions of a space, such as desk location in an office, can be closely associated with certain identities, and hence can be used to deanonymize the users. Therefore, by tracking a pseudonym and gathering related clues (for example, where the pseudonym spends most of its time and whether the pseudonym spends more time at a certain desk than anyone else), the adversary can easily find out the user identity, although the pseudonym is used.

To relieve the threat of linking attack, anonymity-based approaches need to make a pseudonym indiscernible with a number of other pseudonyms. To achieve this, most approaches introduce a trusted intermediary to coordinate users and to provide a large enough anonymity set. In this section we discuss four anonymity-based countermeasures in detail, and at the end of this section, we present a brief comparison of these works.

9.4.1 k-Anonymity

The concept of k-anonymity is originally proposed in [120] in order to provide protection for linking attack. A released data set is considered to be k-anonymous if every element in it is indistinct with at least k-1 other elements. In other words, every combination of values of attributes can be indistinctly matched to at least k elements.

Gruteser and Grunwald [121] extend the k-anonymous concept to the scope of location information. A subject is considered as k-anonymous if and only if the location of the subject is indistinguishable from the locations of at least k-1 other subjects. If a k-anonymous individual reports his location, attackers cannot tell which of the k subjects actually locates at the reported location.

Now the problem turns to be how to achieve k-anonymity. The location information can be represented by a tuple of three intervals ($[x_1, x_2]$, $[y_1, y_2]$, $[t_1, t_2]$). $[x_1, x_2]$ and $[y_1, y_2]$ describe a region in two-dimensional space where the subject is located at a time span $[t_1, t_2]$. Basically, a set of tuples that dissatisfies the k-anonymity requirement can be converted to a k-anonymous set by generalization. Generalization is similar to the degrading techniques used for obfuscation, which decreases the precision of the revealed information. For example, two distinct intervals [12, 23] and [24, 37] can be generalized to [12, 37] and becomes indistinguishable. Since the location information contains both spatial and temporal information, generalization can be applied spatially and/or temporally.

The basic idea of spatial cloaking is to choose a sufficiently large area so that enough number of subjects inhabit this region. Obviously, a larger region means less precision and lower quality of services. Therefore, the challenge is to report spatial information as precise as possible while satisfying the k-anonymity constraint. The algorithm in [121] uses the quadtree to achieve this objective. It keeps dividing an area into quadrants of equal size, until further dividing would create a quadrant with less than k subjects, as illustrated in Fig. 9.1. Each subject reports its host quadrant as its spatial information.

Temporal cloaking, the orthogonal approach to spatial cloaking, tends to reveal more precise spatial coordinates while reducing the precision in time dimension. The idea is to delay a service request containing location information until k individuals have visited the same area of the requestor. Temporal cloaking can be combined with spatial cloaking to make a balance between spatial and temporal resolution.

Certainly, a trusted intermediary is necessary for this approach, since it requires a global knowledge of the distribution of users. If the k-anonymity constraint is satisfied, an attacker only has a probability of $1/k$ at the most to figure out the identity of a user.

Nonetheless, Bettini et al. [122] point out that simple k-anonymity might be insufficient since an attacker can track the historical location information of a pseudonymous user and analyze the movement pattern (e.g., the commuter route of a pseudonym). To mitigate this type of attack, they introduce the notion of

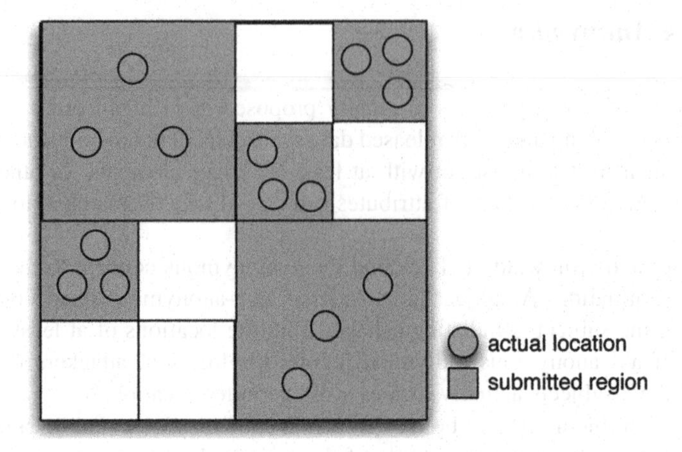

Fig. 9.1 An example of achieving three-anonymity by quadrants dividing

"historical k-anonymity," which concerns that the personal history of location received by a service provider cannot be distinguished from k-1 other sets of personal history of location received by the same service provider.

Generally speaking, spatial and temporal cloaking provides a limited protection for location privacy. Tracking the path of a user can break the protection easily. Also, this approach sacrifices spatial and temporal resolution of location information as well as the quality of service.

9.4.2 Mix Zone

The method of "mix zones" [123] introduces a trusted middleware. A user registers a list of location-based applications that he is interested in with the middleware. An application receives event callbacks about the user from the middleware when the user enters or exits the areas related to this application. The middleware updates user location periodically and issues callbacks to applications when necessary. When communicating with service providers, the middleware uses pseudonyms instead of identities so as to protect privacy.

Mix zone is designed to solve two main drawbacks of anonymity-based approaches. First, it is obvious that the longer a user keeps using a same pseudonym, the weaker the anonymity becomes. The anonymity would be invalidated if the identity of a subject one gets revealed at any location on its path. For example, if a user divulges the identity and location (probably due to the imprudence) in some messages caught previously, then the user appoints a new anonymous message to the middleware. Unfortunately, this measurement does not work. The attacker can link the later message with the previous ones.

Second, the history of location information provides clues that can help attackers figure out the identity of a subject. Suppose an attacker knows that a pseudonym's home and office are in regions A and B, respectively. But the attacker fails to figure out the identity of the pseudonym because there are at least k different pseudonyms in each region. If considering the two clues simultaneously, the attacker might be able to reidentify the pseudonym, since the individual satisfying both constraints (home in region A and office in region B) might be unique.

A direct countermeasure is to change user pseudonym frequently. However, it brings out two new problems. First, some applications might not work properly with fast-changing pseudonyms. Second, if the spatial and temporal resolution provided by the middleware is sufficiently high, attackers can still link the old and new pseudonyms.

To solve the two problems, the concept of "mix zone" is proposed. A mix zone for a group of users is defined as a connected spatial region of maximum size in which none of these users have registered any application callback. The areas where some users have registered for callbacks are called application zones. Users keep using same pseudonyms within the same application zone. When users are inside a mix zone, applications would not receive any location information about them. The following measurement makes the user identities "mixed." When a user enters an application zone from a mix zone (or enters a mix zone from an application zone), the user is assigned with a new, unused pseudonym. As a result, when appears in a mix zone, a user cannot be distinguished from others inside the mix zone at the same time. Also, it is difficult to link a user coming out of a mix zone with any user who enters the mix zone previously.

Figure 9.2 shows an example of this procedure. Suppose there are two users who have registered services in airport, bank, and coffee shop. At some time, one user is in the airport and the other is in the coffee shop. Their presence might be aware by all three service providers since the providers can communicate with each other. Afterwards both users have entered the mix zone and have their pseudonyms changed. When one of the two users enters the bank zone, the service providers only see a new pseudonym appears, but they cannot know which previously appeared pseudonym should be linked to this new pseudonym, since it could be either one of the two users.

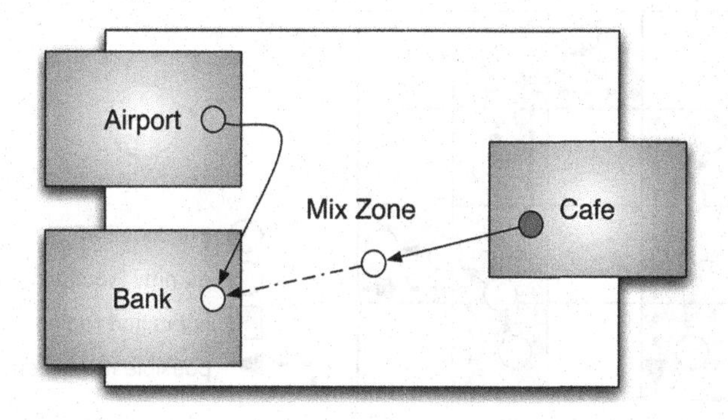

Fig. 9.2 Mix zone example

However, how to divide users into different groups and set the mix zone for these groups is complex. A large mix zone would reduce the security due to the relevance in spatial and temporal coordinates. A user entering a coffee shop in downtown cannot be the one who just appears in the airport one minute ago. A small mix zone increases the difficulty of pseudonym mixing, since it requires the diversity of pseudonyms inside a mix zone.

9.4.3 Using Dummies

Kido et al. propose a way to fool attackers by using dummies [124]. When a user sends position information to a service provider, the report is attached with a set of fake position data which are called "dummies," as illustrated in Fig. 9.3. From the view of the service provider, it looks like there are several different user requests. The provider answers these requests by sending back a message (which contains all the responding to these positions) to the user. The user only selects the necessary data corresponding to his location.

However, if the dummies are generated randomly, observers can easily tell apart the true location and the dummies, because the distance that a subject can move in a fixed time interval is limited. To avoid this, the dummy behavior should be related to the user. Two dummy generalization algorithms are presented in [124]: moving in a neighborhood and moving in a limited neighborhood.

Compared to the k-anonymity approaches, using dummies have several advantages. First, it is difficult for attackers to find out the true pattern of movement of an individual. Second, users can report precise location information with high spatial and temporal resolution, so that little quality of services would be lost. This approach has a drawback that it increases the cost of communication. Users need to report additional dummy location information to service providers, and service providers need to return additional service data for the dummies. Only a small fraction of the communications is useful and all dummy-related communications are overheads.

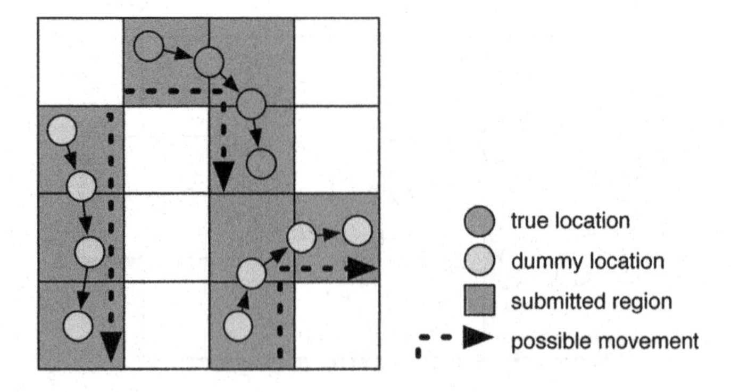

Fig. 9.3 Dummy generation. Attackers cannot determine the true movement

9.4.4 *Path Confusion*

The *k*-anonymity approaches and the mix zone have a common weakness: they all rely on the density of individuals. If the density is not sufficient, the *k*-anonymity approaches deserve poor quality of services due to imprecise location information, while the mix zone might provide poor anonymity since attackers can easily link pseudonyms by temporal and spatial relevance.

Path confusion is proposed to preserve privacy in GPS traces, which can guarantee a certain level of location privacy even for users in low-density regions [125]. The idea is similar to temporal cloaking but it works on paths. The intermediary would delay releasing the user's location, until it finds out the user's path intersects with another user's. Then the intermediary reveals all locations on the two paths altogether, as illustrated in Fig. 9.4. Attackers can only see a bundle of locations on the two paths occurring at the same time. The attacker can tell neither which path the target being tracked is on, nor which direction on the path the target is heading for. Therefore, the target being tracked is confused with other individuals. To provide better anonymity, the intermediary can simply wait longer until more paths are intersected.

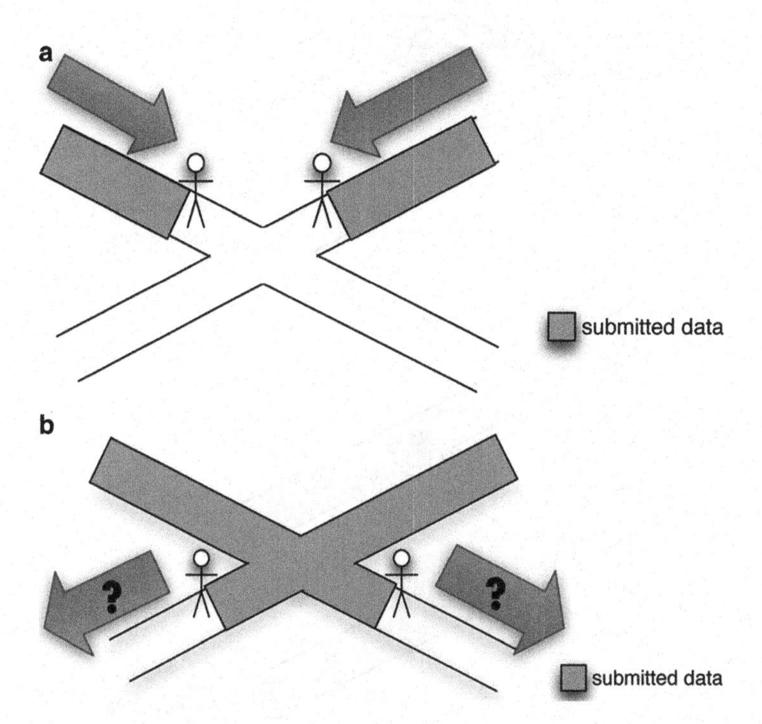

Fig. 9.4 Path confusion. (**a**) At $t = t_0$, an attacker can track the two users according to previously revealed locations. and (**b**) At $t = t_0 + \varepsilon$, since the two users have coincided in space and time, the attacker cannot say whether they turn or go straight

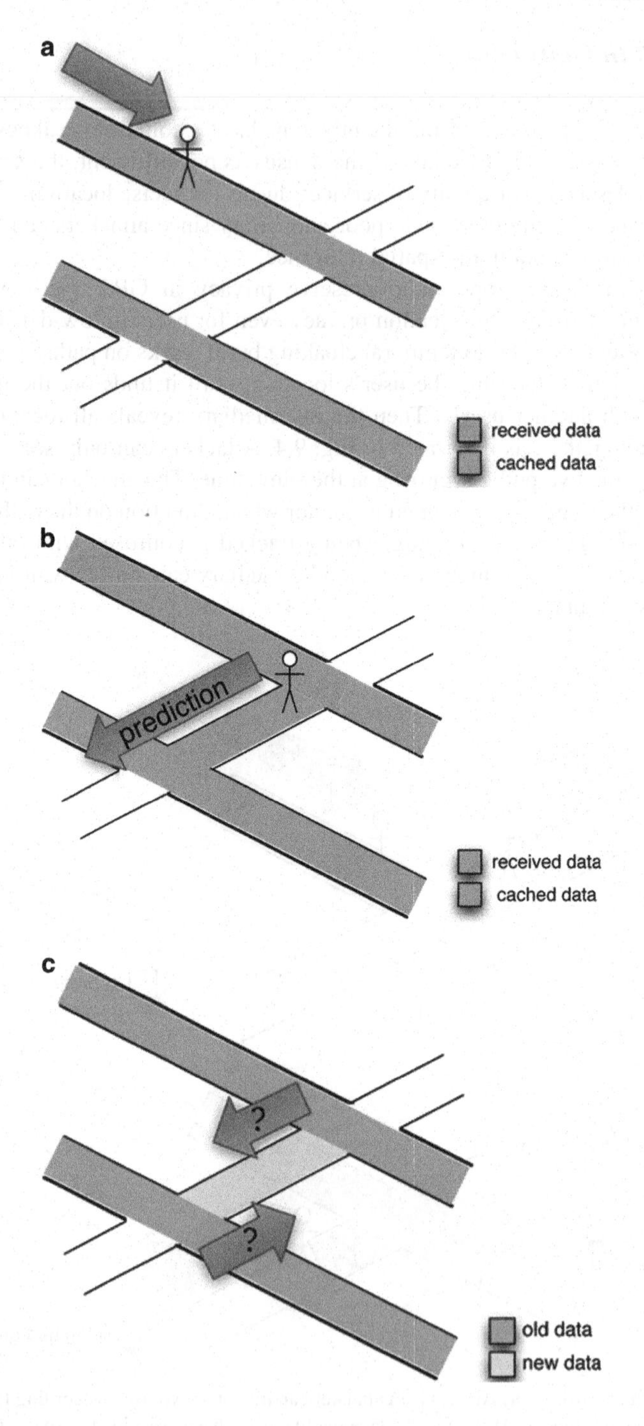
a
received data
cached data
b
prediction
received data
cached data
c
?
?
old data
new data

Releasing precise location information, path confusion can keep the quality of services. The main drawback is that, similar to temporal cloaking, it sacrifices real-time services due to the delay of requests.

Meyerowitz and Choudhury develop cache cloak based on the idea of path confusion [126]. Rather than posteriori analysis of a user's path, cache cloak prefers using mobility prediction to do a prospective form of path confusion. It keeps a spatial cache which contains data for a set of position points. If a user submits a position point that hits the cache, then the intermediary returns the cached data for that location directly, without fetching data from the service provider. If a user submits a position that is not in the cache, which means a cache miss, cache cloak would generate a predicted path for the user. The predicted path is extrapolated until it reaches another path that exists in the cache. (i.e., the predicted path is connected on both ends to other cached paths) The entire predicted path is then submitted to the service provider and all responses for locations along the path are retrieved and cached. Moving along a path, a user gets serviced directly from the cache until deviates from the predicted.

From the attackers' view, each location release contains a bunch of locations on a path. Each newly released path connects two paths released previously, say, path A and path B, as illustrated in Fig. 9.5. There are three possible cases that will trigger a new query: the user on path A turns toward path B; the user on path B turns toward path A; and a new user on the newly released path begins to use the service. Attackers cannot tell apart the three possibilities and accordingly fail to track users.

Cache cloak does not degrade the spatial or temporal resolution as the dummy-based approach. Moreover, a predicted path can be viewed as a dummy (which confuses attackers), and probably this kind of dummies acts more "reasonable" than the dummies generated by the two algorithms, moving in a neighborhood and moving in a limited neighborhood. For the cost of communication, cache cloak does not increase any unnecessary communication between users and the intermediary, although it brings about unnecessary communications between the intermediary and service providers. This overhead can be low if the cache cloak intermediary and service providers are connected by wired networks.

9.4.5 Comparison

Before comparing anonymity-based algorithms, we need to answer the problem that how we can tell if an algorithm is better than any other? The level of location privacy can be reflected by the size of anonymity set, but the definition of anonymity set varies

---

Fig. 9.5 An example for cache cloak. (**a**) A user is moving along a previously cached path. He retrieves data from the intermediary directly. (**b**) A user deviates his path, which triggers a cache miss. New path is predicted and service data along the predicted path are requested from the service providers by the intermediary, and all the retrieved data are stored in the cache. (**c**) An attacker cannot determine what triggers the new data queries. It could be users turning in from the upper street (path A), or from the lower street (path B)

Table 9.1 Anonymity-based approaches

	Loss of QoS	Antitracking	Cost of comm.	Intermediary
Spatial & temporal cloaking	Degraded	Not capable	No increase	Necessary
Mix zone	Degraded	Capable	No increase	Necessary
Dummy	Not affected	Capable	Increased (wireless)	Not necessary
Cache cloak	Not affected	Capable	Increased (wired)	Necessary

among different approaches. In addition, the size of anonymity is usually a parameter that can be set flexibly if necessary. Besides anonymity, we try to characterize an anonymity approach by the following factors:

1. *Loss of quality of service (QoS).* Approaches, such as spatial and temporal cloaking, which degrade the resolution of location information would certainly sacrifice the quality of service. Mix zone breaks the continuity of services, which might degrade the quality of services as well.
2. *Antitracking ability.* We have shown that the historical location data, a.k.a. the pattern of movement, would lead to privacy leaks. Approaches like spatial and temporal cloaking cannot curb an attacker from extracting information through tracking, while some approaches like path confusion can deal well with the attacks based on tracking.
3. *Cost of communication.* Approaches like using dummies and cache cloak increase the cost of communication. The cost of communication on wired network is much cheaper than that on wireless network.
4. *Intermediary dependence.* Most approaches require a trusted intermediary. However, the deployment of an intermediary is expensive, and the communication between users and an intermediary needs to be protected from being interrupted; otherwise, all the efforts would be meaningless.

At last, we summarize the anonymity-based approaches in Table 9.1.

9.5 Summary

Although a lot of approaches have been proposed, a number of issues remain open.

Distributed anonymity. Most anonymity-based approaches require a trusted intermediary, but what if an intermediary cannot be trusted? Or the communication between users and an intermediary is not secure? Dummy-based approach gives a solution without an intermediary, but it increases the cost of communication. Can users cooperate without an intermediary? These questions are still unanswered.

Other types of attacks. Anonymity-based approaches only solve linking attack problem, but are vulnerable for other types of attacks, such as homogeneity attack. Taking k-anonymity as an example, the lack of diversity inside the anonymity set might leak user privacy. For instance, if a location region is inside an abortion clinic or AIDS clinic, in spite of several indistinguishable subjects inside the region, an attacker can still infer the activity of a victim as long as the victim is among these subjects. How to protect privacy from other types of attacks? These problems are worth researching.

Hybrid schemes. No approach can solve the privacy problem perfectly and a combination of privacy strategies might be more effective. How to make different strategies working together is need to be studied.

Pervasive and mobile computing changes the scale of the privacy issue. Future privacy protection approaches are expected to deal with a large number of users, a flood of service requests, and highly frequent data updates. In summary, the privacy issue must be fully addressed before the real proliferation of pervasive computing and the Internet of things (IoT).

ERRATUM TO:

Yunhao Liu • Zheng Yang

Location, Localizaton, and Localizability

Location-awareness Technology for Wireless Network

The original version of this publication unfortunately contained a mistake. Dr. Yunhao Liu's affiliation was incorrect. The corected affiliation is provided below.

Yunhao Liu
Department of Computer Science and Engineering
The Hong Kong University of Science and Technology
Email: Liu@cse.ust.hk

Y. Liu and Z. Yang, *Location, Localization, and Localizability: Location-awareness Technology for Wireless Networks*, DOI 10.1007/978-1-4419-7371-9_10,

References

1. U. Varshney, "Pervasive healthcare," *IEEE Computer*, vol. 36, no. 12, pp. 138–140, 2003
2. G. Borriello, V. Stanford, C. Narayanaswami, W. Menning, "Pervasive computing in healthcare," *IEEE Pervasive Computing*, vol. 6, no. 1, 2007
3. M. Satyanarayanan, "Pervasive computing: vision and challenges," *IEEE Personal Communications*, vol. 8, no. 4, pp. 10–17, 2001
4. M. Weiser, "The computer for the twenty-first century," *Scientific American*, vol. 265, no. 3, pp. 94–104, 1991
5. R. Angeles, "RFID technologies: supply-chain applications and implementation issues," *Information Systems Management*, vol. 22, no. 1, pp. 51–65, 2005
6. R. Glidden, "Design of ultra-low-cost UHF RFID tags for supply chain applications," *IEEE Communications Magazine*, vol. 42, no. 8, 2004
7. Z. Yang, M. Li, Y. Liu, "Sea depth measurement with restricted floating sensors," in Proceedings of IEEE RTSS, 2007
8. M. Li, Y. Liu, "Underground coal mine monitoring with wireless sensor networks," *ACM Transactions on Sensor Networks (TOSN)*, vol. 5, no. 2, 2009
9. L. Mo, Y. He, Y. Liu, J. Zhao, S. Tang, X.-y. Li, G. Dai, "Canopy closure estimates with GreenOrbs: sustainable sensing in the forest," in Proceedings of ACM SenSys, 2009
10. "GreenOrbs Project," in http://greenorbs.org/ 2010
11. G. Tolle, J. Polastre, R. Szewczyk, N. Turner, K. Tu, S. Burgess, D. Gay, P. Buonadonna, W. Hong, T. Dawson, D. Culler, "A macroscope in the redwoods," in Proceedings of ACM SenSys, 2005
12. G. Werner-Allen, K. Lorincz, J. Johnson, J. Lees, M. Welsh, "Fidelity and yield in a volcano monitoring sensor network," in Proceedings of USENIX OSDI, 2007
13. G. Barrenetxea, F. Ingelrest, G. Schaefer, M. Vetterli, O. Couach, M. Parlange, "SensorScope: out-of-the-box environmental monitoring," in Proceedings of ACM/IEEE IPSN, 2008
14. J. Li, J. Jannotti, D.S.J.D. Couto, D.R. Karger, R. Morris, "A scalable location service for geographic ad hoc routing," in Proceedings of ACM MobiCom, 2000
15. G. Kortuem, J. Schneider, D. Preuitt, T.G.C. Thompson, S. Fickas, Z. Segall, "When peer-to-peer comes face-to-face: collaborative peer-to-peer computing in mobile ad hoc networks," in Proceedings of Peer-to-Peer Computing, 2001
16. N. Patwari, J.N. Ash, S. Kyperountas, A.O. Hero, R.L. Moses, N.S. Correal, "Locating the nodes: cooperative localization in wireless sensor networks," *IEEE Signal Processing Magazine*, vol. 22, no. 4, pp. 54–69, 2005
17. A.H. Sayed, A. Tarighat, N. Khajehnouri, "Network-based wireless location: challenges faced in developing techniques for accurate wireless location information," *IEEE Signal Processing Magazine*, vol. 22, no. 4, pp. 24–40, 2005

Y. Liu and Z. Yang, *Location, Localization, and Localizability: Location-awareness Technology for Wireless Networks*, DOI 10.1007/978-1-4419-7371-9,

18. B. Karp, H.T. Kung, "GPSR: greedy perimeter stateless routing for wireless networks," in Proceedings of ACM MobiCom, 2000
19. K. Alzoubi, X.-y. Li, Y. Wang, P.-j. Wan, O. Frieder, "Geometric spanners for wireless ad hoc networks," *IEEE Transactions on Parallel and Distributed Systems (TPDS)*, vol. 14, no. 4, pp. 408–421, 2003
20. N. Li, J.C. Hou, "Localized topology control algorithms for heterogeneous wireless networks," *IEEE/ACM Transactions on Networking (TON)*, vol. 13, no. 6, 2005
21. M. Cardei, D.-Z. Du, "Improving wireless sensor network lifetime through power aware organization," *Wireless Networks*, vol. 11, no. 3, pp. 333–340, 2005
22. P.-J. Wan, C.-W. Yi, "Coverage by randomly deployed wireless sensor networks," *IEEE/ ACM Transactions on Networking (TON)*, vol. 14, no. SI, pp. 2658–2669, 2006
23. F. Xue, P.R. Kumar, "On the θ-coverage and connectivity of large random networks," *IEEE/ ACM Transactions on Networking (TON)*, vol. 14, no. SI, pp. 2289–2299, 2006
24. X.-Y. Li, P.-J. Wan, O. Frieder, "Coverage in wireless ad hoc sensor networks," *IEEE Transactions on Computers*, vol. 52, no. 6, pp. 753–763, 2003
25. Q. Fang, J. Gao, L.J. Guibas, "Locating and bypassing routing holes in sensor networks," in Proceedings of IEEE INFOCOM, 2004
26. A.A. Abbasi, M. Younis, "A survey on clustering algorithms for wireless sensor networks," *Computer Communications*, vol. 30, pp. 2826–2841, 2007
27. G. Mao, B. Fidan, B.D.O. Anderson, "Wireless sensor network localization techniques," *Computer Networks*, vol. 51, pp. 2529–2553, 2007
28. S.Y. Seidel, T.S. Rappaport, "914 MHz path loss prediction models for indoor wireless communications in multifloored buildings," *IEEE Transactions on Antennas and Propagation*, vol. 40, no. 2, pp. 209–217, 1992
29. P. Bahl, V.N. Padmanabhan, "RADAR: an in-building RF-based user location and tracking system," in Proceedings of IEEE INFOCOM, 2000
30. N.B. Priyantha, A. Chakraborty, H. Balakrishnan, "The cricket location-support system," in Proceedings of ACM MobiCom, 2000
31. A. Savvides, C. Han, M.B. Strivastava, "Dynamic fine-grained localization in ad-hoc networks of sensors," in Proceedings of ACM MobiCom, 2001
32. K. Whitehouse, D. Culler, "Calibration as parameter estimation in sensor networks," in Proceedings of ACM WSNA, 2002
33. "Nanotron Technologies," in http://www.nanotron.com
34. C. Peng, G. Shen, Y. Zhang, Y. Li, K. Tan, "BeepBeep: a high accuracy acoustic ranging system using COTS mobile devices," in Proceedings of ACM SenSys, 2007
35. L. Girod, D. Estrin, "Robust range estimation using acoustic and multimodal sensing," in Proceedings of IROS, 2001
36. C. Knapp, G. Carter, "The generalized correlation method for estimation of time delay," *IEEE Transaction on Acoustics, Speech, Signal Processing*, vol. 24, no. 4, pp. 320–327, 1976
37. N.B. Priyantha, A. Miu, H. Balakrishnan, S. Teller, "The cricket compass for context-aware mobile applications," in Proceedings of ACM MobiCom, 2001
38. A. Nasipuri, K. Li, "A directionality based location discovery scheme for wireless sensor networks," in Proceedings of WSNA, 2002
39. L. Doherty, K.S.J. Pister, L.E. Ghaoui, "Convex position estimation in wireless sensor networks," in Proceedings of IEEE INFOCOM, 2001
40. A. Galstyan, B. Krishnamachari, K. Lerman, S. Pattem, "Distributed online localization in sensor networks using a moving target," in Proceedings of ACM/IEEE IPSN, 2002
41. T. He, C. Huang, B.M. Blum, J.A. Stankovic, T.F. Abdelzaher, "Range-free localization schemes in large scale sensor networks," in Proceedings of ACM MobiCom, 2003
42. N. Bulusu, J. Heidemann, D. Estrin, "GPS-less low cost outdoor localization for very small devices," *IEEE Personal Communications Magazine*, vol. 7, no. 5, pp. 28–34, 2000

43. S. Gezici, Z. Tian, G.B. Giannakis, H. Kobayashi, A.F. Molisch, H.V. Poor, Z. Sahinoglu, "Localization via ultra-wideband radios," *IEEE Signal Processing Magazine*, vol. 22, no. 4, pp. 70–84, 2005
44. S. Venkatraman, J. Caffery Jr., "Hybrid TOA/AOA techniques for mobile location in non-line-of-sight environments," in Proceedings of WCNC, 2004
45. A. Catovic, Z. Sahinoglu, "The Cramer-Rao bounds of hybrid TOA/TSS and TDOA/RSS location estimation schemes," *IEEE Communications Letters*, vol. 8, pp. 626–628, 2004
46. L. Cong, W. Zhuang, "Hybrid TDOA/AOA mobile user location for wideband CDMA cellular systems," *IEEE Transactions on Wireless Communications*, vol. 1, no. 3, pp. 439–447, 2002
47. R.I. Reza, "Data fusion for improved TOA/TDOA position determination in wireless systems," Virginia Tech., Master Thesis, 2000
48. M. Youssef, M. Mah, A.K. Agrawala, "Challenges: device-free passive localization for wireless environments," in Proceedings of ACM MobiCom, 2007
49. Y. Liu, L. Chen, J. Pei, Q. Chen, Y. Zhao, "Mining frequent trajectory patterns for activity monitoring using radio frequency tag arrays," in Proceedings of PerCom, 2007
50. W.H. Foy, "Position-location solutions by Taylor-series estimation," *IEEE Transactions on Aerospace and Electronic Systems*, vol. AES-12, pp. 187–194, 1976
51. B. Friedlander, "A passive localization algorithm and its accuracy analysis," *IEEE Journal of Oceanic Engineering*, vol. 12, no. 1, pp. 234–245, 1987
52. Y.T. Chan, K.C. Ho, "A simple and efficient estimator for hyperbolic location," *IEEE Transactions on Signal Processing*, vol. 42, no. 8, pp. 1905–1915, 1994
53. D.J. Torrieri, "Statistical theory of passive location systems," *IEEE Transactions on Aerospace and Electronic Systems*, vol. AES-20, no. 2, pp. 183–198, 1984
54. L.M. Ni, Y. Liu, Y.C. Lau, A. Patil, "LANDMARC: indoor location sensing using active RFID," *ACM Wireless Networks*, vol. 10, no. 6, 2004
55. Y. Shang, W. Ruml, Y. Zhang, M.P.J. Fromherz, "Localization from mere connectivity," in Proceedings of ACM MobiHoc, 2003
56. J. Bachrach, C. Taylor, "Localization in sensor networks," in *Handbook of Sensor Networks: Algorithms and Architectures*, I. Stojmenovic, ed., New York: Wiley, 2005
57. Y. Shang, W. Ruml, "Improved MDS-based localization," in Proceedings of INFOCOM, 2004
58. D. Niculescu, B. Nath, "Ad hoc positioning system (APS)," in Proceedings of IEEE GLOBECOM, 2001
59. D. Niculescu, B. Nath, "DV based positioning in ad hoc networks," *Journal of Telecommunication Systems*, vol. 22, no. 1–4, pp. 267–280, 2003
60. K. Whitehouse, A. Woo, C. Karlof, F. Jiang, D. Culler, "The effects of ranging noise on multi-hop localization: an empirical study," in Proceedings of ACM/IEEE IPSN, 2005
61. D. Niculescu, B. Nath, "Error characteristics of ad hoc positioning systems (APS)," in Proceedings of ACM MobiHoc, 2004
62. D. Moore, J. Leonard, D. Rus, S. Teller, "Robust distributed network localization with noisy range measurements," in Proceedings of ACM SenSys, 2004
63. D. Goldenberg, P. Bihler, M. Cao, J. Fang, B. Anderson, A.S. Morse, Y.R. Yang, "Localization in sparse networks using sweeps," in Proceedings of ACM MobiCom, 2006
64. S. Capkun, M. Hamdi, J.P. Hubaux, "GPS-free positioning in mobile ad hoc networks," in Proceedings of Hawaii International Conference on System Sciences, 2001
65. X. Wang, J. Luo, S. Li, D. Dong, W. Cheng, "Component based localization in sparse wireless ad hoc and sensor networks," in Proceedings of IEEE ICNP, 2008
66. L. Kleinrock, J.A. Silvester, "Optimum transmission radii for packet radio networks or why six is a magic number," in Proceedings of IEEE National Telecommunications Conference, 1978
67. R. Nagpal, H. Shrobe, J. Bachrach, "Organizing a global coordinate system from local information on an ad hoc sensor network," in Proceedings of ACM/IEEE IPSN, 2003

68. H. Lim, J.C. Hou, "Localization for anisotropic sensor networks," in Proceedings of IEEE INFOCOM, 2005
69. Y. Wang, J. Gao, J. Mitchell, "Boundary recognition in sensor networks by topological methods," in Proceedings of ACM MobiCom, 2006
70. S. Lederer, Y. Wang, J. Gao, "Connectivity-based localization of large scale sensor networks with complex shape," in Proceedings of IEEE INFOCOM, 2008
71. Y. Wang, S. Lederer, J. Gao, "Connectivity-based sensor network localization with incremental Delaunay refinement method," in Proceedings of IEEE INFOCOM, 2009
72. Z. Guo, Y. Guo, F. Hong, X. Yang, Y. He, Y. Liu, "Perpendicular intersection: locating wireless sensors with mobile beacon," in Proceedings of IEEE RTSS, 2008
73. Z. Zhong, T. He, "Achieving range-free localization beyond connectivity," in Proceedings of ACM SenSys, 2009
74. R.J. Fontana, S.J. Gunderson, "Ultra-wideband precision asset location system," in Proceedings of Ultra Wideband Systems and Technologies, 2002
75. S. Lanzisera, D. Lin, K. Pister, "RF time of flight ranging for wireless sensor network localization," in Proceedings of WISES, 2006
76. Z. Yang, Y. Liu, "Quality of trilateration: confidence based iterative localization," in Proceedings of IEEE ICDCS, 2008
77. T. Eren, D.K. Goldenberg, W. Whiteley, Y.R. Yang, A.S. Morse, B.D.O. Anderson, P.N. Belhumeur, "Rigidity, computation, and randomization in network localization," in Proceedings of IEEE INFOCOM, 2004
78. H.L.V. Trees, *Detection, estimation and modulation theory, part I*. New York: Wiley, 1968
79. J. Liu, Y. Zhang, F. Zhao, "Robust distributed node localization with error management," in Proceedings of ACM MobiHoc, 2006
80. A. Kannan, B. Fidan, G. Mao, B. Anderson, "Analysis of flip ambiguities in distributed network localization," in Proceedings of Information, Decision and Control, 2007
81. C. Savarese, K. Langendoen, J. Rabaey, "Robust positioning algorithms for distributed ad-hoc wireless sensor networks," in Proceedings of USENIX Annual Technical Conference, 2002
82. D. Liu, P. Ning, W. Du, "Attack-resistant location estimation in sensor networks," in Proceedings of ACM/IEEE IPSN, 2005
83. J. Hwang, T. He, Y. Kim, "Secure localization with phantom node detection," in Proceedings of IEEE INFOCOM, 2007
84. L. Jian, Z. Yang, Y. Liu, "Beyond triangle inequality: sifting noisy and outlier ranging measurements," in Proceedings of IEEE INFOCOM, 2010
85. P.J. Rousseeuw, A.M. Leroy, *Robust regression and outlier detection*. New York: Wiley, 1987
86. Z. Li, W. Trappe, Y. Zhang, B. Nath, "Robust statistical methods for securing wireless localization in sensor networks," in Proceedings of ACM/IEEE IPSN, 2005
87. H.T. Kung, C.-K. Lin, T.-H. Lin, D. Vlah, "Localization with snap-inducing shaped residuals (SISR): coping with errors in measurement," in Proceedings of ACM MobiCom, 2009
88. M. Badoiu, E.D. Demaine, M. Hajiaghayi, P. Indyk, "Low-dimensional embedding with extra information," in Proceedings of ACM Annual Symposium on Computational Geometry (SCG), 2004
89. T. Liu, C. Sadler, P. Zhang, M. Martonosi, "Implementing software on resource-constrained mobile sensors: experiences with Impala and ZebraNet," in Proceedings of ACM MobiSys, 2004
90. J. Luo, D. Wang, Q. Zhang, "Double mobility: coverage of the sea surface with mobile sensor networks," in Proceedings of IEEE INFOCOM, 2009
91. M. Grossglauser, D.N.C. Tse, "Mobility increases the capacity of ad hoc wireless networks," *IEEE/ACM Transactions on Networking*, vol. 10, no. 4, pp. 477–486, 2002
92. S. Capkun, J.-P. Hubaux, L. Buttyan, "Mobility helps security in ad hoc networks," in Proceedings of ACM MobiHoc, 2003
93. Q. Wang, X. Wang, X. Lin, "Mobility increases the connectivity of K-hop clustered wireless networks," in Proceedings of ACM MobiCom, 2009

94. B. Dil, S. Dulman, P. Havinga, "Range-based localization in mobile sensor networks," in Proceedings of EWSN, 2006
95. L. Hu, D. Evans, "Localization for mobile sensor networks," in Proceedings of ACM MobiCom, 2004
96. M. Rudafshani, S. Datta, "Localization in wireless sensor networks," in Proceedings of IEEE/ACM IPSN, 2007
97. S. Datta, C. Klinowski, M. Rudafshani, S. Khaleque, "Distributed localization in static and mobile sensor networks," in Proceedings of WiMob, 2006
98. S. Guha, R. Murty, E.G. Sirer, "Sextant: a unified node and event localization framework using non-convex constraints," in Proceedings of ACM MobiHoc, 2005
99. W. Xi, J. Zhao, X. Liu, X.-Y. Li, Y. Qi, "EUL: an efficient and universal localization method for wireless sensor network," in Proceedings of IEEE ICDCS, 2009
100. J.-g. Park, E.D. Demaine, S. Teller, "Moving-baseline localization," in Proceedings of IEEE/ACM IPSN, 2008
101. B. Hendrickson, "Conditions for unique graph realizations," *SIAM Journal of Computing*, vol. 21, no. 1, pp. 65–84, 1992
102. B. Jackson, T. Jordan, "Connected rigidity matroids and unique realizations of graphs," *Journal of Combinatorial Theory Series B*, vol. 94, no. 1, pp. 1–29, 2005
103. G. Laman, "On graphs and rigidity of plane skeletal structures," *Journal of Engineering Mathematics*, vol. 4, pp. 331–340, 1970
104. J.E. Hopcroft, R.E. Tarjan, "Finding the triconnected components of a graph," Department of Computer Science, Cornell University TR 140, 1972
105. D.J. Jacobs, B. Hendrickson, "An algorithm for two-dimensional rigidity percolation: the pebble game," *Journal of Computational Physics*, vol. 137, pp. 346–365, 1997
106. L. Lovasz, Y. Yemini, "On generic rigidity in the plane," *SIAM Journal on Algebraic and Discrete Methods*, vol. 3, no. 1, pp. 91–98, 1982
107. J. Aspnes, T. Eren, D.K. Goldenberg, A.S. Morse, W. Whiteley, Y.R. Yang, B.D.O. Anderson, P.N. Belhumeur, "A theory of network localization," *IEEE Transactions on Mobile Computing (TMC)*, vol. 5, no. 12, pp. 1663–1678, 2006
108. Z. Yang, Y. Liu, X.-Y. Li, "Beyond trilateration: on the localizability of wireless ad-hoc networks," in Proceedings of IEEE INFOCOM, 2009
109. J.B. Saxe, "Embeddability of weighted graphs in k-space is strongly NP-hard," in Proceedings of Allerton Conference Communication, Control and Computing, 1979
110. D. Goldenberg, A. Krishnamurthy, W. Maness, Y.R. Yang, A. Young, A.S. Morse, A. Savvides, B. Anderson, "Network localization in partially localizable networks," in Proceedings of IEEE INFOCOM, 2005
111. Z. Yang, Y. Liu, "Understanding node localizability of wireless ad-hoc networks," in Proceedings of IEEE INFOCOM, 2010
112. A.F. Westin, *Privacy and freedom*. New York: Atheneum, 1967
113. M. Duckham, L. Kulik, "Location privacy and location-aware computing," in *Dynamic & Mobile GIS: Investigating Change in Space and Time*. Boca Raton: CRC, 2006, pp. 34–51
114. U. S. D. to Justice and O. o. I. a. Privacy, *Overview of the Privacy Act of 1974*, 2004
115. J. Peterson, *A presence-based GEOPRIV location object format*, 2004
116. WWWC (W3C), "Platform for privacy preferences project (p3p)," 2005
117. C.A. Gunter, M.J. May, S.G. Stubblebine, "A formal privacy systems and its application to location-based services," in Proceedings of Workshop on Privacy Enhancing Technologies, 2004
118. IBM, "The enterprise privacy authorization language (EPAL 1.1)," 2005
119. M. Duckham, L. Kulik, "A formal model of obfuscation and negotiation for location privacy," *Pervasive*, pp. 152–170, 2005
120. P. Samarati, "Protecting respondents identities in microdata release," *IEEE Transactions on Knowledge and Data Engineering*, vol. 13, no. 6, pp. 1010–1027, 2001

121. M. Gruteser, D. Grunwald, "Anonymous usage of location-based services through spatial and temporal cloaking," in Proceedings of ACM MobiSys, 2003
122. C. Bettini, X.S. Wang, S. Jajodia, "Protecting privacy against location-based personal identification," *Secure Data Management*, pp. 185–199, 2005
123. A.R. Beresford, F. Stajano, "Location privacy in pervasive computing," *IEEE Pervasive Computing*, vol. 2, no. 1, pp. 46–55, 2003
124. H. Kido, Y. Yanagisawa, T. Satoh, "An anonymous communication technique using dummies for location-based services," in Proceedings of ICPS, 2005
125. B. Hoh, M. Gruteser, H. Xiong, A. Alrabady, "Preserving privacy in gps traces via uncertainty-aware path cloaking," in Proceedings of ACM CCS, 2007
126. J. Meyerowitz, R.R. Choudhury, "Hiding stars with fireworks: location privacy through camouflage," in Proceedings of ACM MobiCom, 2009

Index